Prophecy, Alchemy, and the End of Time

Prophecy, Alchemy, and the End of Time

John of Rupescissa in the Late Middle Ages

Leah DeVun

COLUMBIA UNIVERSITY PRESS
NEW YORK

Columbia University Press
Publishers Since 1893
New York   Chichester, West Sussex
Copyright © 2009 Columbia University Press
Paperback edition, 2014
All rights reserved

Library of Congress Cataloging-in-Publication Data
DeVun, Leah.
Prophecy, alchemy, and the end of time : John of Rupescissa in the late Middle Ages / Leah DeVun.
    p. cm.
Includes bibliographical references and index.
ISBN 978-0-231-14538-1 (cloth : alk. paper)—
ISBN 978-0-231-14539-8 (pbk. : alk. paper)—
ISBN 978-0-231-51934-2 (e-book)
1. Johannes, de Rupescissa, ca. 1300–ca. 1365. 2. Alchemy—Religious aspects—Christianity—History—To 1500. 3. Religion and science—Europe—History—To 1500. 4. Pharmacology—Europe—History—To 1500. 5. Apocalyptic literature—History and criticism. I. Title
[DNLM: 1. Johannes, de Rupescissa, ca. 1300–ca. 1365. 2. Religion and Science. 3. Alchemy. 4. History, Medieval. BL 245 D514p   2009]
BR115.A57D48   2009
274´.05—dc22   2008030108

Columbia University Press books are printed on permanent and durable acid-free paper.
Printed in the United States of America

Cover image: *Time Life Pictures*/Getty Images
Cover design: Milenda Nan Ok Lee

For my parents, and for Mom

CONTENTS

List of Illustrations
ix

Acknowledgments
xi

1. Introduction
1

2. The Proving of Christendom
11

3. John of Rupescissa's Vision of the End
32

4. Alchemy in Theory and Practice
52

5. Artists and the Art
80

6. Metaphor and Alchemy
102

CONTENTS

7. The End of Nature
129

8. Conclusion
150

Notes
165

Bibliography
223

Index
245

ILLUSTRATIONS

Figure 1. Portrait of an Angelic Pope from the *Ascende calve* prophecy
25

Figure 2. Tree of life watered by Christ's blood
121

Figure 3. Christ's blood fertilizes poppies beneath the crucifix
122

Figure 4. The Zodiac Man
131

ACKNOWLEDGMENTS

In the years that it has taken to finish this book, I've enjoyed the support of many generous colleagues, friends, and family members, without whom this project would have been impossible. I owe special thanks to the staff at the Huntington Library, Columbia University's Butler Library, and Texas A&M University's Evans Library, especially to Bob Scott, Consuelo Dutschke, Karen Green, and Joel Kitchens. I would also like to thank the institutions that provided me with support through fellowships and awards, including the Huntington Library, the UCLA Center for Medieval and Renaissance Studies, the University of Kansas Jerry Stannard International Memorial Award Committee, and the Texas A&M University's Melbern G. Glasscock Center for Humanities Research, College of Liberal Arts, Women and Gender Studies Program, and History Department. I thank everyone involved with these institutions for their financial and practical assistance. I would also like to express my gratitude to the Institute for Research in the Humanities at the University of Wisconsin-Madison. A fellowship from the institute allowed me to finish this manuscript; its staff and fellows offered me both friendship and a lively and rigorous academic atmosphere.

I've been fortunate to work with a number of inspiring professors, who have shaped this project as well as my path as a scholar. In particular, I'd like to thank Matthew Jones, Adam Kosto, Nancy Siraisi, and Robert Somerville, all of whom worked with me on my dissertation. Robert Lerner has seen me through this project from the beginning, when he suggested that I take a closer look at an obscure apocalyptic prophet named John of Rupescissa. I'm extremely grateful for the time he has spent reading my manuscripts and for the many tips and discoveries he has passed my way. Laura

ACKNOWLEDGMENTS

Smoller and Joel Kaye have been unbelievably generous in their willingness to read my chapters and in their enthusiasm for my work. I also would like to thank the many scholars with whom I have discussed this project over the years, including Renate Blumenfeld-Kosinski, Lisa Cooper, Rachel Fulton, Jeffrey Hamburger, Marlene Hennessey, Bruce Holsinger, T. K. Hunter, Richard Keyser, Richard Kieckhefer, Michael McVaugh, Barbara Newman, William Newman, Tara Nummedal, Nicole Rice, Michael Ryan, Thomas Safely, David Sorkin, Katja Vehlow, and Lee Wandel. I apologize in advance if I have omitted anyone. My undergraduate professors at the University of Washington, Robert and Robin Stacey, Mary O'Neil, Dauril Alden, and Robert Tracy McKenzie, also deserve thanks for passing along their love of history and pointing me toward a career in academia. Most of all, I thank my mentor, Caroline Walker Bynum, whose intelligence, originality, kindness, and tireless dedication to her students long beyond their graduation have inspired me—and meant more to me than I can express here. It has been a pleasure and an honor to work with her. She made this book—and so much more—a reality.

My fantastic colleagues at Texas A&M have contributed so much to this volume, and I am grateful to every member of the History Department for encouraging my scholarship and making College Station a welcoming place. I also appreciate their understanding for the semesters that I was away on fellowships, and I thank my department head, Walter Buenger, and the university administration for facilitating my leaves. Of my A&M colleagues, Daniel Bornstein, Donnalee Dox, Cary Nederman, and James Rosenheim, in particular, have been kind enough to read and offer their opinions of portions of this manuscript. I also thank the support and secretarial staff in the History Department and the Vice President for Research offices who helped me on countless occasions.

My editor at Columbia University, Wendy Lochner, my copyeditor, Michael Haskell, and an anonymous reader also played a significant role in shaping this book. Wendy deserves special thanks for championing my manuscript and for guiding me through the publication process.

Anna Trumbore, Anna Harrison, and Mary Doyno have been the most brilliant and sympathetic friends that anyone could hope for. They read many drafts of my book, helped me with professional and personal hurdles, and reminded me often and in so many ways why I began this career. Other friends, including Roy Allen, D'Lane Compton, Jennifer and Tony Elias, Heather Hagle, Frank Howard, Rachel Howe, Tonya Huskey, Derek

ACKNOWLEDGMENTS

Marks, Jill Mycka, K. Naca, Philip Nickel, Coya Paz, Nina Rubin, Nicole Russell, Kristen Schilt, Stephanie Soileau, Larin Sullivan, Purvi Shah, Astria Suparak, Vicky Tamaru, Megan Wright, and Madeline Yale all made sure that I had a life outside of my work. I must especially thank Ian Sullivan, whose presence through the years has meant so much to me.

Finally, I would like to thank my parents, Gail and Esmond DeVun, for their encouragement in this and all other things, as well as my extended family of grandmothers, uncles, aunts, and cousins. I am especially grateful to my partner, Macauley Elmer, for her support and love, which helped me to see the light at the end of all of this.

Prophecy, Alchemy, and the End of Time

INTRODUCTION

From the confines of his prison cell in the mid-fourteenth century, the Franciscan friar and alchemist John of Rupescissa boldly predicted the destruction of the society that held him captive. Rupescissa (also called Jean de Roquetaillade or Rochetaillade) was born around 1310 near the town of Aurillac, in the region of southern France known as the Auvergne.[1] By the age of roughly forty-six, when he wrote his *Liber ostensor quod adesse festinant tempora* (The Book That Shows That the Times Are Soon to Be at Hand), he was a tireless predictor of the apocalypse and a thorn in the side of the church.[2] By 1356, Rupescissa had already been imprisoned for more than a decade: he was bound in chains, locked under a staircase in the hopes he might die quietly, even declared a *fantasticus*—a foolish or insane person—by the highest juridical body in the Christian church, the papal court at Avignon. But all of this poor treatment in no way deterred Rupescissa from spreading his message. The last days were coming; the horrors of Antichrist were about to begin.

Although some members of the high clergy supported Rupescissa, his apocalyptic tendencies were hardly welcomed by the majority of his superiors in the Franciscan order and the larger church. During a hearing before the papal curia in 1354, Rupescissa's examiner was openly skeptical, ridiculing Rupescissa and throwing his books to the ground. According to Rupescissa's account in the *Liber ostensor*, the judges asked how he could presume to know the future when the many learned doctors of the church did not. The friar answered defiantly that a person had to be "whitened" and "dyed" in order to understand divine secrets.[3] The years of persecution and imprisonment he suffered at the hands of the church, Rupescissa argued, actually primed him to receive supernatural gifts from God. He explained:

INTRODUCTION

Those who have been obligated to capture secrets [in the past] had to be cooked in the fire of many tribulations; therefore, [as Isaiah] says, "and just like fire they were proved." For nearly twelve years now I have been cooked in prison cells, and the fiery boilings do not cease, but continually burn hotter against me.... And therefore it pleases God to reveal to me—"fantastic" and insane—the secret.[4]

Rupescissa went on to claim that he had secret knowledge of the events of the apocalyptic crisis, including the dates of apocalyptic floods, plagues, wars, and, ultimately, the appearance of Antichrist himself. He also offered innovative and highly inflammatory predictions about the identity of Antichrist and the duration of the millennium that would follow his reign.

Rupescissa's obsession with apocalypticism had roots in the tragedies of his lifetime. The Great Famine, the Black Death, and the Hundred Years War between England and France were terrible events that transformed Western Europe and prompted intense spiritual reflection among observers. The apocalyptic scenes described in Christian prophecy, particularly those in St. John's Apocalypse, appeared hauntingly similar to contemporary events. Apocalyptic spirituality thus became for many, including Rupescissa, a way to understand the natural and political disasters they witnessed. For Rupescissa, it also gave meaning to his own personal disasters—his humiliating arrest and many years of imprisonment. He came to view the events of his lifetime as an apocalyptic storm with his own trials at its center. The dramatic episodes of his day were predicted by astrology and prophecy; his persecution paralleled that of the martyrs and prophets of the past.

But Rupescissa's statement to the curia at his 1354 hearing hints at something more than his penchant for apocalypticism and self-aggrandizement: his language of boiling, cooking, and proving also evokes the traditional process of an alchemical transmutation. According to Rupescissa, humans who wished to understand prophetic secrets were tested by the heat of flames and transformed by whitening and dying. Purgatorial cleansing by fire was a standard biblical image, but Rupescissa's language brings to mind the alchemical purifications so important to two of his other writings, the *Liber lucis* (Book of Light, c. 1350) and the *Liber de consideratione quinte essentie omnium rerum* (Book on the Consideration of the Quintessence of All Things, 1351–1352), both manuals aimed at combating apocalyptic disasters through alchemy.[5] The images that Rupescissa selected to describe his life in the *Liber ostensor*—the cooking, the whitening, the proving—were thus highly significant, for alchemy was central to Rupescissa's apocalyptic message. Rupescissa claimed that humans could weather the apocalypse by using metals and medicines made with alchemy. Furthermore, Rupescissa found in Christian eschatology ideas that helped him both to understand alchemy and to explain it to his readers through metaphors borrowed from religion.

Much as Rupescissa lamented the scant attention paid to his recommendations by church authorities in the fourteenth century, modern scholars might now lament how few published works on Rupescissa have appeared in recent decades, making him one of the most underappreciated figures in the history of science. Rupescissa has been credited by scholars such as Robert P. Multhauf and Robert Halleux, among others, with the invention of "medical chemistry," a genre of early modern alchemical literature devoted to human health, now considered the conceptual forerunner of pharmacology.[6] It is clear that Rupescissa was responsible for at least two significant innovations: first, he invented the "quintessence," an extremely influential alcohol-based medicine that overturned conventional wisdom about the human body and the cosmos. Second, he seems to have been the first to argue that a vital, heavenly principle could be distilled from different herbs, animal products, and minerals (particularly gold, mercury, vitriol, and antimony) to produce different internal medicines suited to different diseases. While alchemy was used for medicinal purposes (particularly external medicine) at least as early as the thirteenth century, Rupescissa's systematic treatise on the application of inorganic materials to internal medicine more closely resembles texts of a much later period.[7] In fact, Rupescissa's theories had enormous consequences for the development of alchemical and *materia medica* tracts in the early modern period. Classic alchemical and distillation works of the sixteenth century, including those by Hieronymous Brunschwygk, Philipp Ulstad, Paracelsus, and Andreas Libavius borrowed heavily from Rupescissa's ideas on distillation and the quintessence and extended them in new directions.[8] Historians of science are now especially interested in these early modern alchemists, whose experiments they view as crucial to the development of modern chemistry. Yet many alchemical concepts and techniques traditionally attributed to early modern alchemists—for instance, the production of medicinal mineral acids and the use of gold in circulated alcohol—already appear in Rupescissa's texts.

But Rupescissa's significance is not only his influence on later generations. His program of alchemy and prophecy allows us insight into the world of the fourteenth century. While modern historians interpret theology, natural philosophy, and medicine, among other subjects, as distinct disciplines, Rupescissa's intellectual system suggests that such boundaries were not shared—or at least not shared in the same way—by those who lived in the late Middle Ages. Rupescissa's alchemy was a religious endeavor and a thoroughly Christian science. His writings situated Greco-Roman natural philosophy and Arab alchemy squarely within the context of Christian apocalyptic time and history. According to Rupescissa, humanity's knowledge of alchemy would reach its apogee during the period of the Christian last days, in which alchemy would play a crucial role that would legitimate its study. In addition, alchemy could intervene in the cosmic events of the apocalyptic crisis through the reenactment of the central episodes of Christian history: the incarnation, the crucifixion, and the resurrection. Rupescissa's synthetic vision of Christian alchemy offers historians a new way of viewing both religion and naturalism. His multifaceted approach to the problem of the apocalyptic crisis—encompassing prophetic exegesis, metal transmutation, astrology, and alchemical medicine—should surely give pause to those who would assume the stability of such disciplinary categories across time.

Rupescissa also offers an important perspective on the attitudes of those who encountered the uncertainties of the late Middle Ages. Rupescissa's writings emphasize the power of human agency to affect the course of history. Rupescissa imagined a version of the last days in which human beings could use prophecy and naturalist knowledge to actively engage with and successfully battle Antichrist and the disasters that would accompany his reign. Rupescissa was apparently the first to argue that humankind could survive this apocalyptic crisis and populate the postapocalyptic millennium, which would endure for a literal one thousand years. In addition, he claimed that such survivors could use the art of alchemy to access the perfection of the superlunary sphere and channel its heavenly effects onto earth. The catastrophes of Rupescissa's lifetime could well have inspired a pessimistic outlook in which individual action was deemed fruitless. Yet for Rupescissa, disaster entailed an obligation for Christians to fight using every possible weapon in their arsenal. Furthermore, he believed that in such a battle they would be victorious.

Rupescissa's alchemical and apocalyptic innovations, and how he came to imagine them, are the focus of this study. Guided by predictions of the end in biblical and "Joachite" prophecy, Rupescissa devised a plan for human beings to escape what he called the "infelicities of nature"—that is, disease, aging, and even death—through alchemy. Rupescissa's immersion in apocalyptic spirituality led him to believe that human bodies could be perfected by a process of purgation and redemption that mirrored apocalyptic prophetic narratives. According to his vision, everything on earth contained within it a "quintessence," a kernel of heavenly perfection that could be distilled and used as a medicine, improving human bodies to the point of near-immortality. His claim constituted a challenge to the most basic tenets of medieval medicine and natural philosophy. Despite his iconoclastic approach to received wisdom about the human body and the universe, contemporary scholastic discussions of medicine and natural philosophy deeply influenced Rupescissa's understanding of natural processes—and how they might be overturned through alchemy.

In recent decades, our knowledge of alchemy has been greatly enhanced by research in the history of science. In response to earlier scholarship that judged alchemy to be a "pseudoscience," a hindrance to the progress of chemistry not worthy of serious study, a new generation of scholars has called for a reevaluation of alchemy because of its significance to premodern religious, intellectual, and cultural trends. Excellent work has been done on individual alchemists or alchemical texts by historians such as Chiara Crisciani, Michela Pereira, William Newman, and Barbara Obrist, who have attempted to place specific theorists within a broader history of medicine and natural philosophy.[9] Recent studies have also explored the mutually beneficial relationship between alchemy and artisanal traditions, which promoted both technological innovation and the documentation of oral traditions.[10] Scholars of early modern alchemy such as Betty Jo Teeter Dobbs and Lawrence Principe have similarly attempted to recast alchemy as crucial to the thought of such iconic figures as Isaac Newton and Robert Boyle, rather than as a misstep in a teleological history of science.[11]

Such studies have made it possible to assess John of Rupescissa's historical context, as well as his contribution to the discipline of alchemy. As I noted, scholars such as Multhauf and Halleux have deemed Rupescissa the originator of "medical chemistry," and they have linked his ideas to the emergence of Paracelsian therapeutics in the sixteenth century. Their work is invaluable for situating Rupescissa within the history of alchemy and medicine, yet they have tended to dismiss Rupescissa's religious writings as unrelated to his scientific ideas. As a result, they have failed to notice much of the complexity of his alchemy, which derived from the different intellectual and spiritual traditions that informed his theory.[12] Historians of religion, such as Robert E. Lerner, Jeanne Bignami-Odier, Marjorie Reeves, and André Vauchez, have similarly produced admirable studies that place Rupescissa's prophecy within a long-standing tradition of apocalyptic hermeneutics.[13] The growing number of studies of Rupescissa's apocalypticism over the last two decades has signaled a resurging interest in his life and legacy. But these analyses have not considered his alchemical writings, and thus they have not addressed the centrality of nature to his reading of salvation history. This book will put John of Rupescissa back together: rather than citing him within separate histories of science and religion, it will evaluate the different aspects of naturalist and religious thought that operate within his work. My purpose is to show that spirituality and naturalism need not be discussed in isolation from each other. Only an integrated approach can account for the multiple bodies of knowledge that intersect with one another and inflect one another in Rupescissa's writings.[14]

Because I intend this book to be accessible to those interested in the history of science and the history of Christianity, I discuss medieval naturalist and apocalyptic concepts in detail in order to demonstrate the originality and significance of Rupescissa's ideas. Some of the material that I include will be familiar to specialists, although because my approach and my emphasis are new, so too are my conclusions. In chapters 2 and 3, I place Rupescissa's life and work within the context of various late-antique and medieval apocalyptic traditions, and I demonstrate how his eschatological views arose from developments in the "Franciscan Spiritual" conflict of the thirteenth and early fourteenth centuries. During this period, the apocalyptic writings of Joachim of Fiore (c. 1135–1202) inspired great enthusiasm among Franciscan friars, who expected the appearance of Antichrist and the establishment of a "third state," an earthly paradise that would follow the apocalyptic crisis and preface the end of time. Joachite and Franciscan Spiritual prophecy suggested an active role for human intervention in the apocalyptic drama by opposing Antichrist and shaping the future millennial age. Rupescissa found in this textual tradition evidence both that the crisis was near and that humankind had an opportunity (and an obligation) to participate in a historic battle between Christ and Antichrist. These assumptions led Rupescissa to turn to alchemy as a solution to the apocalyptic disasters at hand.

INTRODUCTION

In chapters 4 and 5, I analyze Rupescissa's alchemical theory, particularly his idea of the "quintessence." Chapter 4 demonstrates the ways in which Rupescissa drew from Aristotelian natural philosophy and Galenic medicine in order to formulate the quintessence, an alchemical remedy that challenged conventional assumptions about both the human body and the order of nature. According to Rupescissa, the quintessence obeyed the rules of a higher level of "perfected nature," which also regulated the celestial bodies of the superlunary sphere. Alchemy thus allowed a piece of "heaven" to enter earth, healing disease and prolonging the lives of corruptible human bodies. Chapter 5 enriches this picture by considering the writings attributed to three well-known naturalist authors, Roger Bacon (1214–1292), Arnald of Vilanova (c. 1240–1311), and Ramón Llull (c. 1232–1316). My discussion shows how each author constructed the science of alchemy as it was known to Rupescissa. This genealogy of sources reconstructs the development of the quintessence and establishes the novelty of Rupescissa's alchemical thought.

Chapter 6 focuses closely on the language of Rupescissa's two alchemical manuals, the *Liber lucis* and *De consideratione quinte essentie omnium rerum* (hereafter, *De quinta essentia*). I evaluate the context and consequences of Rupescissa's use of code names, metaphors, and similes (invoking heaven, Christ, the crucifixion, and the resurrection, among other images) to describe alchemical processes. I examine Rupescissa's choice of imagery alongside images in other texts of the period, including those of theological doctrine and devotional literature. I argue that in such figurative language we are able to see the deep structural relationship between Rupescissa's apocalyptic and alchemical systems. This section of the book represents a new approach to the much-studied topic of alchemical imagery. My focus upon Rupescissa's imagery allows us to see alchemy neither in terms of Jungian symbol nor as a precursor to chemistry but rather as an element of fourteenth-century religious reform.

Finally, chapter 7 focuses on alchemy and astrology as naturalist technologies to intervene in salvation history. According to Rupescissa, nature was destined to play a central role in the apocalyptic crisis to come. Through astrology, human beings could read the "Book of Nature" to discover hidden apocalyptic prophecy; through alchemy, they could raise the corrupted state of terrestrial nature to a higher level of perfection, paving the way for a millennial state on earth. Rupescissa's assertion that the Book of Nature and the Book of Scripture were parallel

{7}

sources of religious truth constitutes an early version of a doctrine that was to become commonplace in the natural theology of the seventeenth, eighteenth, and nineteenth centuries. The origins of this view are often located by scholars in the revival of Neoplatonism and Hermetism during the Renaissance by humanists such as Marsilio Ficino. Yet Rupescissa's sentiments—expressed a century earlier—point to the development of early modern approaches to God and nature in the apocalyptic context of the late Middle Ages.

Because religion was such a pervasive aspect of medieval life, historians of science have generally dismissed religious language in alchemy as not particularly significant. As a result, scholars have not studied carefully the specific religious terminology that is so obtrusive and characteristic a part of medieval alchemy. In contrast, this study will demonstrate that such language was critical to the genesis and communication of naturalist ideas during the Middle Ages. Scholars of modern science now believe that metaphors are an inescapable means by which scientists and their audiences conceive and construct the world.[15] Modern scientific metaphors such as "greenhouse gases" or "electron orbits" are so familiar to us that they may at first seem uncomplicated in their meaning. Yet these metaphorical systems are central to the way in which scientists (and consequently the general public) perceive the world in which we live. Rupescissa's religious metaphors are similarly important: the language he chose shaped the way in which he and his readers viewed the natural world and the human body. Religious thinking was at the heart of Rupescissa's alchemy and, as I shall argue, at the heart of a foundational moment in the history of science.

Not only did religion shape Rupescissa's alchemy, but alchemy also influenced Rupescissa's religious activities. Alchemy provided a model for his apocalypticism, molding his prophetic exegesis and providing a foundation for his predictions. Alchemy's processes were comparable to Rupescissa's method as a prognosticator: in order to make his predictions, Rupescissa collected the base materials of prophecy and scripture, distilled their content through careful exegesis, and extracted their most pure and precious essence, which he recorded for readers. Such a method dovetails well with the alchemical opus that Rupescissa described and promoted in his writings—a series of secret and purgative operations aimed at the production of precious metals and improved human health. Like the alchemical distillation process, the process of exegesis refined materials until their most significant element could be secured and directed towards the needs of the Christian community.

And Rupescissa used the metaphor of alchemy to describe his life. He, too, was an alchemical product, cooked and tested by trials, made worthy of divine secrets by fire. Such metaphors are emblematic of Rupescissa's analytic process. He applied the metaphorical imagery of religion freely to alchemy, drawing parallels that inspired his theories and ultimately transformed the discipline. Analogical thinking led to Rupescissa's vision of alchemy as an apocalyptic melding of heaven and earth, a process that could transmute corrupt earthly matter into a perfection mirroring the heaven described in Scripture. Rupescissa applied the language and methods of prophecy and alchemy to each other, allowing them to cross-pollinate. Such a combination proved fertile, leading Rupescissa to develop a new genre of alchemical writing, as well as an important set of naturalist concepts that had a legacy in the Middle Ages and beyond.

Rupescissa's revolutionary formulation of alchemy and apocalypticism was by no means an obscure episode in medieval history. His ideas were well known and well received both in his own time and in later centuries, as demonstrated by the large number of extant copies of his work, which date from the fourteenth to the eighteenth centuries. Robert Halleux has identified 55 manuscript copies of Rupescissa's *Liber lucis* and more than 140 manuscript copies of his *De quinta essentia*; both also appear in numerous printed editions.[16] *De quinta essentia* also received perhaps the highest compliment from one member of its audience: a late-fourteenth-century reader plagiarized large portions of the text, inserting its ideas into the widely read *De secretis nature*, which eventually became attached to the name of Ramón Llull.

Rupescissa's prophetic writings were similarly popular: the *Vade mecum in tribulatione* (1356), a short summary of apocalyptic predictions, exists in English, French, Catalan, Castilian, Italian, German, and Czech versions. A number of these translations appeared within decades of the *Vade mecum*'s first appearance in Latin, indicating the wide and rapid dissemination of Rupescissa's ideas. Famous during his lifetime, Rupescissa and his travails were documented by a number of chroniclers of the mid-fourteenth century. Moreover, later redactors of his prophecies continued to produce updated versions of the texts to reflect new historical circumstances, demonstrating the continuing appeal and relevance of Rupescissa's apocalyptic rhetoric. The large number of manuscripts and editions of both Rupescissa's alchemical and prophetic texts firmly establishes the importance and lasting influence of his writings.

The story of John of Rupescissa begins with his tragic life, but there is another, larger story here, one that extends into many directions of medieval and early modern history. Through Rupescissa's work, we may see the incorporation of Greek and Arab philosophy into Christian thought, the interaction of scholastic and nonscholastic intellectual traditions, the conflict over wealth and authority within the Church and the Franciscan order, the cultural impact of the natural and political disasters of the fourteenth century, and the fusion of medicine and alchemy into early modern pharmacology. Because Rupescissa's interests and activities touched upon many of the most pressing and exciting topics of his day, his writings offer historians a means to appreciate the overlap and interplay of these different stories.

During his imprisonment, Rupescissa wrote fervently to prepare readers for the grim future he imagined: the wars, famines, and plagues on the horizon. Along the way, he created a new genre of alchemical writing and a new cosmology of heaven and earth. Rupescissa saw in alchemy microcosmic narratives of Christian history, in which miniature cycles of violence, death, and resurrection could both reenact and act upon a larger narrative of human redemption. He saw in prophecy and astrological horoscopes explanations of the uncontrollable, even incomprehensible events that he and his contemporaries experienced. The lines of spiritual and naturalist thought of the fourteenth century provided to Rupescissa a comfort of sorts: they gave meaning to suffering, and they suggested that this world contained within it a potential for change—whether biological, chemical, social, or political—that could transform earthly life into something more pure and lasting. The conclusions to which Rupescissa came are demonstrative of the power of medieval apocalyptic and naturalist traditions to make sense of human struggle and to foster a hope for transcendence. In their strength and flexibility, these two traditions underpinned Rupescissa's ideas and, in their combination, they provided the spark of the new that transformed them.

THE PROVING OF CHRISTENDOM

In 1356, John of Rupescissa warned in his prophetic treatise, the *Vade mecum in tribulatione*, that within just a few short years, Antichrist would appear among Christians, ushering in a terrifying climax of human history:

Before we come to the year 1365, there will appear publicly an eastern Antichrist whose disciples will preach in parts of Jerusalem with false signs and portents to bring about the seduction of all with error ... in the five years between 1360 and 1365, destructions will abound beyond all human estimation: tempests never before seen from the sky, floods of water unheard of in many parts of the world (except for the Great Flood), serious famines all over the world, plagues and mortalities, outpourings of blood and other corruptions of the passions, through the wounds of which the greatest, weighty part of the present generation will be killed.[1]

As the passage indicates, only a minority of humankind could expect to survive the assault of Antichrist, a horror that Rupescissa depicts as nearly unimaginable. But the passage also tells us something else, something crucial to Rupescissa's approach to the crisis. He describes not only what was to happen—storms, floods, plagues, and famines—but also precisely *when* they would happen. Like many medieval writers, Rupescissa argued that the events of the end were in some sense "beyond all human estimation"—yet much could be known about them in advance. Clues to these last mysterious events were purportedly hidden in prophetic Scripture, particularly in the biblical books of Ezekiel, Isaiah, Daniel, and the Apocalypse of St. John, as well as in extrascriptural revelations by saintly figures and pseudonymous prophets. With the right interpretive method, human beings could ferret out the meaning of these texts and prepare for the *what* and *when* of the last days.

Rupescissa's thoughts on the end fit within a long-standing medieval tradition of "last things," that is, discourse about the last or furthest times of human and cosmic existence.[2] Last things meant not only the end of an individual life, but also the collective end of human existence on earth, as described in Jewish and Christian prophecy. Rupescissa, like many before him, identified the events of his own day with the events of the last days, drawing upon an essentially unchanging apocalyptic narrative that could be applied to changing historical circumstances. Rupescissa was, of course, not the first to use such apocalyptic strategies. Particularly important to his writings was the apocalyptic theology of Joachim of Fiore (c. 1135–1202), as well as the texts that arose from the Spiritual Franciscan controversy of the thirteenth and early fourteenth centuries, including those by the theologian Peter Olivi (1248–1298). These apocalyptic texts provided Rupescissa with the foundational materials for his work. Joachim of Fiore's new apocalyptic chronology suggested an active role for human intervention in salvation history in order to bring about a "third state," a final paradisical age on earth. This scenario was taken up by thirteenth- and fourteenth-century Franciscan friars to contextualize a conflict over poverty within the Franciscan order and to identify Franciscan Spiritual partisans as apocalyptic heroes. Rupescissa was particularly drawn to these writings, and he incorporated their methods and prophecies into his own works. First, he obtained an interpretive framework, which helped to make meaningful the disasters of his day by placing them within the context of a religious teleology. Second, he developed a powerful apocalyptic rhetoric, which aided the presentation of his naturalist program of alchemy, medicine, and astrology as a solution to the problem of Antichrist. Finally, he built upon the thought of previous apocalypticists to create a significant role for human action in the context of the last days.

The future crisis envisioned by Rupescissa is probably familiar to readers as the "apocalypse," as we tend to call it in modern speech, but the language historians use for the end is more complicated. Writings such as those of Rupescissa are categorized as not only "apocalyptic," but also "eschatological" and, sometimes, "chiliastic," "millennial," or "millenarian."[3] Historians use the term "eschatology"—from the Greek word *"eschatos,"* meaning last or furthest—for discourse about what I have described above as "last things."

"Apocalypticism"—from the Greek *"apokalypsis,"* meaning revelation—is used for any assertion that the end, however conceived, is imminent, and that it will unveil the meaning of history. Some scholars have distinguished between the term "apocalypse," which technically refers to a genre of revelatory literature, and "apocalypticism" or "apocalyptic spirituality," which refer to a more general teleological view of history that appears in many forms of literature.[4] Descriptions of the end in medieval texts are often accompanied by those of a new beginning. Many Christians expected a perfected terrestrial society in the future (based on Apocalypse 20:4–6's allusion to a one-thousand-year reign of Christ and the saints on earth). Scholars generally use the terms "millennialism," "millenarianism," and "chiliasm" interchangeably to refer to this expectation of a supernaturally instituted improvement of life on earth before the last judgment and eschaton, regardless of its duration.

Scholars of modern millennialism have arrived at some defining characteristics of this belief throughout Christian history. Joseph F. Zygmunt, who has worked on twentieth-century Jehovah's Witnesses, provides a particularly cogent summary:

[Chiliastic] belief systems represent a curious blend of escapist and quasi-revolutionary orientations, well conveyed in their central convictions: that the prevailing social order is doomed to more or less imminent destruction; that it will be replaced by an ideal system from which all evil will be banished; and that this cataclysmic change will be effected, not by human effort, but by some supernatural agency.[5]

Zygmunt's observations provide a useful starting point for an analysis of Rupescissa's thought. The fourteenth century was a fertile site for the application of preexisting apocalyptic hermeneutics—what the historian Robert E. Lerner has characterized as a "basically unchanging medieval prophetic structure"—to a series of contemporary events.[6] To make sense of the natural and social disasters of the mid-fourteenth century, John of Rupescissa looked to a continuous yet flexible tradition of apocalypticism, which he used to shape both a predictive history and a course for human action. Although Rupescissa borrowed from a wide range of texts, his most important sources derived from two distinct apocalyptic strains: the Sibylline and Joachite narratives of the end.[7] These strains were themselves variations on an exegetical tradition that stretched back to the first days of the institutional church.

THE PROVING OF CHRISTENDOM

PATRISTIC INTERPRETATIONS OF LAST THINGS

The Apocalypse of St. John, along with its antecedents among the traditional Jewish apocalypses, provided much of the basis for early and medieval Christian interpretations of last things. Christian theology holds that time is linear rather than cyclical, and that human beings live within a historical process whose beginning and end are instituted by God. Theorists such as Frank Kermode have proposed that human beings have imagined this structure in order to give meaning to what would otherwise be an intolerable existence "in the middest."[8] If only people could locate their position in relation to beginning and end—and they generally located it very near the end—they could create for themselves a more satisfying sense of participation in the narrative of history.

Such a narrative can be pieced together from passages of the Old and New Testaments: the vivid and highly symbolic descriptions of the end in Daniel, Matthew, and especially St. John's Apocalypse lent themselves to scrutiny and supported a wide variety of readings. The problem of interpreting the last days was among the first to engage early Christian exegetes, and it became increasingly important as the hoped-for *parousia*, or second coming of Christ, failed to materialize. Theologians who lived before the Christian emperor Constantine's reign viewed St. John's Apocalypse as a literal description of the contemporary church rocked by imperial persecutions in the dawn of the second coming.[9] But once the church made peace with the temporal institutions of the post-Constantinian empire, the search for signs of an imminent end among current events became less relevant and more problematic.

The Bible also warned against such interpretations. Acts 1:7 was frequently cited as evidence that knowledge of last things was outside of the bounds of human inquiry. When Augustine sought to disarm millennial expectations in the early fifth century, he invoked Acts in his *On the City of God*, urging contemporaries to refrain from their calculations:

Yet some have said that four hundred, some five hundred, others even a thousand years may be reached between the Lord's ascension and his last coming. But to show how each of them supports his opinion would take too long, and is not necessary; for they use human speculations, and produce nothing decisive from the authority of canonical scriptures. Truly he relaxes the fingers of all who do sums on them about this matter and bids them be at rest, when he says: 'It is not yours to know the times that the Father has fixed by his own power' (Acts 1:7)."[10]

Augustine attempted to shift the climax of history from the second coming to the incarnation, thereby deflating the millennial hopes of his contemporaries. His interpretation of apocalyptic Scripture was allegorical rather than literal: he identified the one-thousand-year-long millennium as symbolic, indicating a spiritual state in the present rather than a material state in the future. The burden thus fell to the individual to secure paradise in the next world rather than to wait for a millennial paradise to be visited upon this world. This orientation of last things toward the "vertical" (emphasizing the individual journey to the other world), rather than the "horizontal" (emphasizing the unveiling of history to a collective humanity) became dominant throughout much of the Middle Ages. Expectations for a millennial age within time were largely discredited by Augustine's analysis: any writer who clung to millennial belief or attempted to predict the second coming was forced to contend with the weight of the saint's opinion. Only in the eleventh century did widespread speculation about the millennial process of history and the collective fate of the church reemerge.[11]

But not all apocalypticists heeded Augustine and ceased to determine the end. A group of prophecies from the East, including the "Tiburtine Sibyl" and the *Revelations* of Pseudo-Methodius, continued apocalyptic speculation outside of the purview of Augustine's proscriptions.[12] Although ostensibly pagan in origin, the Sibylline oracles were afforded great respect by the early Christian Fathers, and they continued to find Christian editors and readers throughout the Middle Ages.[13] The prophecy of Pseudo-Methodius, a seventh-century text attributed to the fourth-century martyr Methodius of Patara, is closely related to the Tiburtine prophecy and shares much of its content. This pair of prophecies is especially notable for its fusion of political and apocalyptic elements. Pseudo-Methodius has been credited with popularizing the "Last World Emperor," an apocalyptic political figure who also appears in Rupescissa's works.

The Sibylline and Pseudo-Methodian prophecies highlight the role of human action in the period preceding a final battle with Antichrist. In the "Tiburtine Sibyl," the Last Roman Emperor, a human leader, engineers a state of peace in the last days of Christendom by conquering the pagans, converting the Jews, and ruling over a bountiful world.[14] But when Antichrist and his armies initiate their attack, the emperor surrenders his crown at

Jerusalem and awaits aid from Christ. Divine intervention is ultimately necessary to slay Antichrist and bring about the eschaton. This formulation—what scholars have described as the "Sibylline" or "pre-Antichrist" tradition—describes a paradise *before* Antichrist and emphasizes human impotence in the face of the crisis. This is in contrast to the "Joachite" or "post-Antichrist" tradition, which places the paradise *after* Antichrist, a distinction that I shall discuss in more detail below. The role of human agency in the last days is a recurrent problem within the prophetic tradition and it becomes central to Rupescissa's apocalyptic program.

JOACHIM OF FIORE AND THE THIRD STATE

With the appearance of Joachim of Fiore, calculators of the end received a whole new set of numbers and dates to activate their fingers. If Augustine pointed to the first half of Acts 1:7 ("It is not yours to know"), Joachim focused on the second half ("the times that the Father has fixed by his own power") and interpreted it as license to explore what signs had already been revealed to humans in scripture.[15] Born in Calabria around 1135, Joachim entered the Benedictine order in 1171.[16] When Cistercianism swept through Italy with its promise of a new and more austere interpretation of the Benedictine rule, Joachim (by then an abbot) petitioned to switch the affiliation of his monastery to the new order. But Joachim became disillusioned with Cistercian life, too; he ultimately founded his own religious order, the Order of Fiore, in an attempt to reform monasticism and prepare for the new age that he anticipated. Famed for his predictions during his lifetime, Joachim was a force in both royal and papal courts, where he fulfilled a number of requests to interpret prophecies and predict the future. Joachim's conflict with the Cistercian order—along with the entanglement of his writings in controversies over the nature of the Trinity and the "Eternal Evangel"—led to his somewhat mixed reputation in the centuries that followed his death. But the myth of Joachim's gift of prophecy, along with the large number of genuine and spurious writings attributed to him, assured him great popularity in the later Middle Ages and a continued influence even in the sixteenth and seventeenth centuries. Joachim left behind a number of treatises in which he claimed to understand the events of the past, present, and future through typological and historical "concordances" (*concordiae*) in scripture. Concordances were symbols or events in the Old and New Testaments that pointed to one another and ahead to John's Apocalypse. In these, Joachim believed the whole history of the church was contained. By noting these concordances, Joachim was able to establish patterns of twos, threes, and sevens that organized history into sequential and overlapping periods. These patterns also allowed for predictions of the end of time.

Much scholarship concerning Joachim considers the orthodoxy of his doctrine, as well as its potential to incite revolutionary action. While it is not necessary for the purposes of this study to grasp the complexities of Joachim's patterns of history, it is essential to understand how they shaped the apocalyptic ideas of the Franciscan order and, consequently, of John of Rupescissa. Although Joachim divided history into a set of overlapping periods of twos and sevens, the crown of his thought (and the element that was most often picked up by his disciples after the 1240s) was his trinitarian interpretation of history. Joachim divided all of history into three overlapping states (*status*): the state of the Father, the state of the Son, and the state of the Holy Spirit. Joachim predicted that during the third and final state, two new religious orders would form—one of contemplatives and one of preachers.[17] Humankind would receive widespread mystical understanding of divine truth and would establish a peaceful and blissful society on earth. The doctrine of the third state was sufficiently complicated that Joachim included explanatory *figurae*, or figurative illustrations, alongside his text. Manuscript copies of the *Liber figurarum* often include images that visualize the three states as linked rings. The rings demonstrate the correspondences among the three states, the Old and New Testaments of the Bible, and the persons of the Trinity.[18]

Joachim's assertion of a period of earthly progression after Antichrist but before the end of time renewed the conversation about the millennium that Augustine had attempted to silence centuries before. Historians have clarified that an independent tradition asserting a "post-Antichrist" peace had continued to develop during the interval between Augustine and Joachim: Jerome, Bede, and Hildegard of Bingen, among others, all allowed for a period of peace (often forty-five days) between the time of Antichrist and the last judgment.[19] But Joachim was the most popular proponent of the idea, and he is generally credited with undermining a purely allegorical interpretation of the millennium. Joachim did not believe that human beings could achieve complete perfection before the end, however. He predicted that a last battle with evil (in the form of the final Antichrist and the tribes of Gog and Magog) would follow the period of peace and immediately precede the eschaton.

Although Joachim was careful not to assign precise dates for the arrival of the third state, his writings led subsequent readers to think that the reign of Frederick II and the period between 1200 and 1260 would be critical. When Frederick II died without revealing himself to be Antichrist and 1260 passed without incident, Joachim's followers were able to postpone the significant date to 1290, then 1335, then later.[20] Joachim's two new orders were also identified by his readers with the two newly formed mendicant orders, who thus appeared primed to play central roles in an apocalyptic near future. Joachim's legacy was to become especially attractive to the Franciscan order by the end of the thirteenth century, when it would become a rallying cry for discontented friars who viewed the order's policies on evangelical poverty through an eschatological lens.

THE FRANCISCAN ORDER AND THE PROBLEM OF POVERTY

The controversy that shook the Franciscan order at the turn of the fourteenth century was rooted in the ideals of its founder, St. Francis of Assisi (1182–1226), who dedicated himself to a life of uncompromising poverty and itinerant preaching in imitation of Christ and the apostles.[21] After his own conversion, Francis persuaded a small group of followers to renounce their material possessions and embrace a radical form of the *vita apostolica*, or apostolic life, characterized by poverty, simplicity, and itinerancy. The first Franciscans—who numbered eleven in the early days of the order—were under the direct guidance of Francis himself and therefore in little need of a comprehensive set of written regulations. Francis, however, envisioned a larger movement that would eventually require some form of organization. Around 1210, he requested papal approval for his order and composed a simple rule that consisted mainly of scriptural quotations.

The force of Francis's personality and the example of his poverty soon attracted thousands of converts to the new order. Francis composed a new rule in 1221 (known as the *Regula Prima* or *Regula Non Bullata*), which clarified the original intent of the order in light of its growth. Further developments soon required further clarifications, and at a chapter meeting in 1222 Francis was importuned to revise the rule again. Under pressure from Franciscan ministers, he relaxed some of the more rigorous restrictions placed upon the brothers. The new rule (the *Regula Bullata*) was confirmed by Pope Honorius III's *Solet annuere* in 1223. In the last days of his life, Francis once more addressed the needs of the growing order with his *Testament*, an addendum to the rule that expressed nostalgia for the early days of the order and urged the brothers to adhere to absolute poverty. Francis's struggle to reconcile his ideal of apostolic life with the practicalities of a growing organization points to the tension within the order that was to lead to great controversy in the decades to come.

Shortly after Francis's death in 1226, the size of the order increased dramatically from its original twelve members to more than 10,000 in Italy alone.[22] As the order expanded, its friars began to fill a variety of roles in church and secular society that appeared to be at odds with the spirit of Francis's rule and *Testament*. The rule posed a particular problem for those trying to adapt the order to its new conditions because Francis forbade glosses of his words. Nevertheless, in 1230 Pope Gregory IX offered an interpretation of the rule intended to address the order's new needs. His *Quo elongati* relaxed the rule's demand of absolute poverty in such a way that it allowed brothers to live in houses and build libraries. He also released Franciscans from obedience to the *Testament*, which he deemed not legally binding. Additional interpretations and relaxations followed: local disputes about the rule were forwarded to a panel of provincial theologians, who subjected the document to legal analysis and made recommendations. Pope Innocent IV's *Ordinem vestrum* (1245) built upon *Quo elongati* to weaken strictures on the brothers' ability to use money.[23] Disputes over the interpretation of the rule became important as donations and legacies inundated the order and Franciscan leaders embarked upon a number of building projects. When Franciscans began to take on positions in universities, royal administrations, and even the papal curia, a number of friars became troubled by the increasing divergence of Franciscan life from its original ideal of poverty.[24] They believed that the order's financial success was fundamentally incompatible with an apostolic vocation as Francis had conceived it. Despite the efforts of the compromise-minded minister-general and theologian Bonaventure (1257–1274) to diffuse the conflict during his tenure, the problem proved intractable. The simmering controversy erupted near the end of the thirteenth century, and the order split into two hostile and opposing factions.

The majority faction, known as the Community or Conventuals, supported relaxations in the observance of the rule to encourage flexibility and stability within the order. Under their leadership, individuals practiced personal poverty, but the order possessed a modified form of corporate wealth: friars were free to use goods that were technically owned by the Roman church. A minority faction of brothers known as the Spirituals, however, argued that increasing corruption and worldliness among Franciscans should be remedied by more rigorous observance of the vow of poverty, which they believed obliged *usus pauper*, or "poor use"—that is, Franciscans should own nothing, and they should use goods only when absolutely necessary.

The leaders of the Spiritual Franciscans seized upon Joachim of Fiore's apocalyptic theology to legitimate their position on *usus pauper*. They saw in Joachim's prophecy a portrait of themselves as members of the new religious orders—and of their persecutors as the minions of Antichrist. The theologian Peter Olivi, who during the last decade of his life became the defender of the *usus pauper* and the unofficial spokesman for the Spiritual position, played a pivotal role in this connection between the Spiritual-Conventual conflict and the Joachite apocalyptic.[25] Olivi's early commentaries on Matthew, Job, and Isaiah, as well as his more famous *Lectura super Apocalypsim* drew heavily upon the works of Joachim.[26] Olivi adopted Joachim's double set of seven ages (*etates*) of world history and placed his own time at the transition between the fifth and sixth periods of the New Testament—a time of moral decline, the reign of Antichrist, and the persecution of a small spiritual elite. He also borrowed Joachim's division of history into three overlapping states, which he juxtaposed in such a way that the sixth and seventh ages corresponded with the beginning of the third state. Olivi furthermore inserted the issue of poverty into the Joachite schema. According to his analysis, poverty and prophecy were mutually dependent; the prophetic meaning of Scripture could be grasped more fully by those who practiced evangelical poverty.[27] In the third state (inaugurated by Francis and soon coming to fruition), an order of "spiritual men" (whom Olivi identified as followers of Francis's rule) would usher in the final church, characterized by poverty and widespread understanding of holy Scripture and divine truth.[28]

Olivi remained in good standing with ecclesiastical authorities until the end of his life (in spite of a condemnation by a Paris commission in 1283 unrelated to his apocalypticism), but a storm followed his death. First, a Franciscan general-chapter meeting in 1299 forbade the reading or owning of Olivi's writings. Next, in 1319, his works were found worthy of condemnation by a papal commission (though no formal condemnation was issued). In the same year the Franciscan general-chapter meeting at Marseilles damned Olivi's errors and ordered all friars who held his books excommunicated. John XXII followed with his own condemnation of Olivi's Revelation commentary in 1326. By the early fourteenth century the leaders of the order and the church had come to view *usus pauper* as a major disciplinary problem and Franciscan apocalypticism as potentially dangerous.[29] Marjorie Reeves has argued that Joachim's trinitarian interpretation of history, taken out of context, could incite subversive action for those who saw themselves at the edge of the third state, a time of flux within temporal institutions and authorities, including the church.[30] Reeves notes three characteristics of such radicalism within Franciscan thought, including "a sense that the extreme crisis of history is about to break upon the world; a belief in the supreme mission of the Order to match this moment; [and] an attitude towards the papacy and the ecclesiastical hierarchy in which obedience strives with the conviction that the Order holds the key to the future, which cannot be wrested from it."[31] The perceived arrogance and anticlericalism of the Spirituals disgusted church authorities, leading John XXII in 1318 to condemn their assertions as "partly heretical, partly insane, and partly [ridiculous]."[32] Particularly disturbing to church officials was an extreme variation on the Spiritual position claiming that the "carnal church" (that is, the Roman church) was to soon be replaced by a spiritual church (led by the Spiritual Franciscans) that would endure until the end of time.[33]

The belief that the Spiritual-Conventual conflict had apocalyptic significance was not confined to the order. The Languedocian and Catalan lay followers of the Spirituals, known as Beguins, began to revere Peter Olivi as their patron saint shortly after his death. Olivi's tomb at Narbonne became such a popular pilgrimage site that the number of visitors was rumored to rival those at St. Francis's Portiuncula.[34] This cultic activity led to the exhumation of Olivi's remains, which were reportedly burned and discarded in the Rhône in 1318 by the order of John XXII. Nevertheless, vernacular copies of Olivi's writings were read aloud at Beguin gatherings, and Beguins who fell into the hands of the Inquisition repeated Spiritual positions on the imminent dawn of a new age, the triumph of the evangelical poor, and the readiness of true Christians to suffer for their convictions.[35] A number of Beguins interrogated by the Inquisition expressed almost fanatical belief in the spiritual men who would institute the third state, a period characterized by the ascendancy of St. Francis's rule over previous dispensations.[36] Because the Spirituals and their followers believed unfalteringly in the rectitude and the urgency of their mission, they were unwilling to submit to any ecclesiastical authority, even the papacy, since they believed temporal institutions had no power to stop the coming apocalyptic age.

A further threat also resulted from an apocalyptic interpretation of the Franciscan order based on Joachim's writings. Although Olivi followed Joachim by declining to name actual people as Antichrists or to give precise dates for apocalyptic events, his disciples were less restrained. Spiritual Franciscans and their followers quickly updated the Joachite and Olivian schemas by adding specific names and dates, some of them highly inflammatory to the ecclesiastical hierarchy. Arnald of Vilanova outraged theologians at Paris in 1300 with his *De tempore adventus Antichristi*, which predicted the eschaton would arrive in 1378.[37] Not only was the institutional church no longer the supreme embodiment of Christ on earth, it might actually be the embodiment of Antichrist. In the *Arbor vitae crucifixae Jesu* (1305), Ubertino of Casale suggested that the mystical Antichrist already ruled in the form of the last two popes, Boniface VIII and Benedict XI.[38] Beguins interrogated by the inquisitor Bernard Gui in the 1320s consistently identified John XXII as the mystical Antichrist. Because the Spirituals and their followers understood the rule of St. Francis to be the gospel, they refused to believe that any pope who attempted to modify or abrogate the rule could be the true pope; instead, he had to be a pseudo-pope or some other manifestation of Antichrist. This conclusion became increasingly attractive to the Spirituals: it gave their struggle with church leaders an apocalyptic significance, justifying their position on poverty (which seemed to align them clearly with Christ's forces) and making them less susceptible to correction (by those seemingly allied with Antichrist).[39] An apocalyptic reading of the conflict moreover compelled Spirituals to take a decisive stand against what appeared to be the ultimate personification of evil: an Antichrist pope. Poverty was no longer a matter of academic debate confined to the order. Devotion to *usus pauper* and Olivi had become a sign that identified its proponents with the forces of good and obligated them to fight the forces of evil.

The interregnum that followed the deaths of Clement V and Franciscan minister-general Alexander of Alexandria in 1314 had allowed the Spiritual movement to grow unimpeded by institutional directives for two years.

With the election in 1316 of John XXII and minister-general Michael of Cesena, however, the church and order redoubled their efforts to put down the rebellion. Pope John XXII issued a flurry of bulls condemning the Spirituals' position and ordering the destruction of Beguin communities in Narbonne and Béziers.[40] In 1317, his *Quorundam exigit* ordered radical Franciscans to adopt appropriate habits and maintain food reserves; *Sancta romana* authorized repressive action against Beguins, Fraticelli, and all lay practitioners of evangelical poverty.[41] In 1318, *Gloriosam ecclesiam* further denounced the Franciscan rebels, urging ecclesiastics and laymen to turn them over to their Franciscan superiors. A delegation of Spirituals from the Midi were imprisoned and four of them burned at the stake at Marseilles the same year. Next, the pope withdrew papal dominion over Franciscan goods with 1322's *Ad conditorem*, forcing Franciscans to accept ownership of possessions rather than continuing to use goods technically owned by the Roman church. Most important, he denied the absolute poverty of Christ and the apostles with 1323's *Cum inter nonnullos*. The Inquisition dealt with recalcitrant Beguins in southern France and Catalonia, culminating in a long series of auto-da-fés during the 1320s.[42] By the end of the decade, the most visible arm of the Spiritual movement had been crushed by a campaign of intimidation and violence.

THE PROVING OF JOHN OF RUPESCISSA

Although the Spiritual cause was greatly weakened by the end of the 1320s, the influence of Spiritual politics and eschatology in the intellectual arena was far from over. Controversy over Olivi, evangelical poverty, and apocalyptic speculation continued to have broad appeal for much of the fourteenth century. John of Rupescissa thus began his vocation in a climate of both long-standing dissatisfaction with church policy among Franciscans and a heightened expectation of radical change in the future. It is unclear for what specific offense Rupescissa was imprisoned in 1344, but he doubtless fell into a tradition of suspect Franciscan ideology at a time when the church was highly alert to potential heirs of the Spiritual rebels. Furthermore, a number of new prophetic writings had appeared in the decades preceding Rupescissa's career to fan the flames of apocalyptic expectation. Illuminated prophecies such as the *Genus nequam* and *Ascende calve* series offered a new set of prophecies centered on the papacy. They included arresting illustrations of Antichrist and angelic popes that made it all the easier for readers to envision a central role for the pope in the coming crisis. (See figure 1 for an image of a heavenly pope attended by an angel.) Rupescissa knew both *Genus nequam* and *Ascende calve* firsthand, and he even wrote a commentary on the latter, now lost.[43] Writings by Spirituals (such as Ubertino of Casale and Angelo Clareno) and their sympathizers (such as Arnald of Vilanova) also continued to circulate, fueling further discussion of the apocalyptic roles to be played by the Franciscan order and the papacy.

John of Rupescissa's life fits well within this Franciscan tradition of apocalypticism and rebellion. In 1332, John of Rupescissa joined the Franciscan order, some five years after he began his studies at the University of Toulouse. Rupescissa probably first encountered speculation about Antichrist in Toulouse: Joachite apocalypticism was particularly popular among the Franciscans there, and Rupescissa was almost certainly exposed to such views at the university.[44] According to his account in the *Liber ostensor*, soon after entering the Franciscan order, Rupescissa experienced a dramatic vision. He was transported in his sleep to the eastern city of Zayton (a city of which he had never heard before), where he spoke to the young Antichrist, who had recently been born in the area. In his vision, Rupescissa preached against Antichrist alongside a group of Franciscans in Zayton then traveled to Rome, where he was finally silenced by Antichrist in a violent confrontation.[45] Rupescissa at first dismissed the vision as a mere dream, but he was surprised to learn several years later that Zayton was a real place (the name refers to Quanzhou, in the Fujian province of China). Once he became convinced of the truth of his dream, Rupescissa fervently believed that God intended him to reveal the content of his vision to the public.

Rupescissa dared to share his visionary insights first with fellow Franciscans at the university around 1335. He publicly predicted an outbreak of war between England and France; he boasted in a later letter that although the pronouncement was ridiculed at the time, its accuracy won him supporters in later years.[46] Later, he publicized his apocalyptic visions more widely, and he may be the same unnamed friar who aired concerns about Antichrist at a Franciscan general chapter meeting in 1337.[47] Sometime between 1337 and 1340, after completing his studies at Toulouse, Rupescissa returned to Aurillac, where he remained at the Franciscan cloister until his arrest several years later. On December 2, 1344, for reasons that are unclear, Rupescissa was removed from the cloister at Aurillac and transported to a Franciscan prison in Figeac. It is likely that Rupescissa's

Figure 1. Portrait of an Angelic Pope from the *Ascende calve* prophecy. *Source:* St. Gall, Kantonsbibliothek (Vadiana), MS 342, p. 13.

sensational predictions finally provoked Franciscan authorities to act, but no record of his offense has survived. At Figeac, Rupescissa was unceremoniously thrown in the mud, initiating his more than twenty years in prison, during which he experienced a version of the apocalyptic horrors he was to predict for all of humanity in his later writings.[48] In his first year of captivity, Rupescissa was restrained in irons and granted little contact with his Franciscan captors. Even worse, in February of 1345, he broke his leg, causing him to undergo a particularly cruel procedure to set the fracture:

Two or three times [I was] placed on a press [*torculari*] to fix the twisted leg, and I was pulled violently to the most extreme torments, just like the martyrs of antiquity. My limbs were pulled until they were nearly pulled out. Thereafter, I stayed for almost one hundred days on a pallet, where no one saw fit to move me or even visit me, except the two or three times a day when a brother brought some wretched food to me. And there I stayed, alone and abandoned, shut up like a dog, and covered in my own bodily waste. I tried as much as I was able to raise the leg that I had broken. Besides that, for sixty days the pestiferous bed that I laid on was not changed once. When this happened, a great multitude of maggots swarmed around my broken leg, which I collected with both hands and threw away.[49]

What pained him even more, Rupescissa added, was that the Franciscan provincial minister, Guillaume Farinier, forbade the friars from hearing his confession, even when he appeared to be near death. After Rupescissa recovered, he was transferred, "ill, weak, and weeping," to various convent prisons in southern France.[50] Finally, in July of 1346, he was brought to Toulouse for a trial before the Dominican inquisitor Jean Dumoulin, a meeting occasioned by accusations (allegedly by Guillaume Farinier) that Rupescissa espoused heretical ideas. Rupescissa claims that despite suspicions about his orthodoxy, he was able to affirm correctly all of the articles of the Catholic faith under examination, leading the inquisitor to order him freed under penalty of excommunication.[51]

Rupescissa's liberty was short-lived, however. The following week, Guillaume Farinier intervened again, sending a letter to the vicar of the convent of Toulouse ordering Rupescissa's rearrest and imprisonment there. A year later, Rupescissa was transferred to the "foul and miserable prison" of Rieux during the feast of St. Peter in Chains.[52] At Rieux, Rupescissa was locked in a small space under the stairs of the dormitory, where he was seized with a serious "malady of the chest" that lasted an entire year. In 1348, having just recovered from his long illness, Rupescissa was infected with the plague, leading to another brush with death. Despite his misfortunes, he obtained little sympathy from the friars of Rieux. Rupescissa remembers that he was "destitute of help and service, so that through the whole day there was no one who would minister to me food, or who would look in on me to see if I was still alive."[53] Rupescissa languished in this way, nearly dead and full of despair, until 1349. On August 1 of that year, an angel appeared to him in his cell, telling him that he would be freed in two weeks, that is, on the anniversary of the day he had first entered the prison.

At that time, the provincial minister of Aquitaine, Raoul de Cornac, fortuitously ordered the transfer of Rupescissa from Rieux to the prison of Castres. But Rupescissa's escort seems to have sympathized with him: instead of taking him to Castres, he allowed Rupescissa to escape to Avignon to plead his case before Pope Clement VI. Rupescissa arrived at the curia in August of 1349 and appeared in consistory on October 2. A sensational hearing ensued, during which Rupescissa defended his apocalyptic predictions and passionately condemned the excesses of the high clergy. According to one report, Rupescissa roused his listeners to such an extent that a priest in the audience decried the corruption of the see of Rome and threw a bull at the head of the pope.[54] Rupescissa's judges may have been hesitant to execute a man who appeared to at least some to be a genuine prophet: at the conclusion of his trial a year later, Rupescissa was neither found guilty of heresy nor released. A compromise allowed him to be declared "fantastic" (*fantasticus*) rather than heretical (*hereticus*), and he was remanded to prison.[55] Despite the apparent conclusion of proceedings against him, Rupescissa was reexamined intermittently and never ceased to fear for his life.

Following his appearance before the curia on October 2, Rupescissa was placed in a papal prison (called the "Sultan's" prison) rather than returned to the custody of the Franciscan order. Rupescissa's hopes for better conditions under the protection of the pope were unfounded, however, for life in the papal prison was in some respects worse than his earlier captivity. In 1351, Rupescissa was joined by Simon Legat, an English priest who would prove to be Rupescissa's nemesis for the next five years. During this period, Rupescissa—with his usual flourish for drama—speculated that "all the saints who were from the beginning of the world until now [have] not sustained as many verbal injuries as I alone suffered night and day in five years from that demoniac [i.e., Legat]."[56]

Legat was chained for six months in a cell near that of Rupescissa, where he goaded the hapless friar, chanting "heretic, tic, tic, heretic" innumerable times until Rupescissa was near despair. Legat insulted Rupescissa continually, calling him "hypocritical" and "insane" and interrupting him whenever he tried to pray, study, or sleep. Even worse, Legat hurled not only insults but also his own feces:

He invented for me a new form of stoning, [Rupescissa writes]: he had a pan which was full of his excrement, and the place where I was studying had a little door next to me. Through that opening he stoned me from time to time; one particular day, with caution I was sitting and studying, and he threw his pan full of excrement against the wall and, rebounding, it fell on me.[57]

Not satisfied with merely annoying Rupescissa, Legat eventually attempted to murder him. Rupescissa remembers:

In the year of our lord 1355, or a little after, in summer, on a certain night, as God had sent me a good almsgiving in food and I had given a part of it to an innocent man who had been unjustly incarcerated. My demoniac [Legat] hid himself beneath the little portal that through which food is passed to the prisoners. He had made a shank of bone, very sharp, like an awl of iron. As soon as I passed my hand through the gate with my food in order to hand them to the unfortunate prisoner, the English traitor threw himself on me without a sound, seizing me by the top of my hood with one hand, and hitting me with the other hand, which held the sharpened shank, with great force. And I would have been killed if the prisoner, to whom I had given the alms, had not put in his hand and by the order of Christ, delivered me from death, and that hand remained perforated and carries a scar still to this day.[58]

These travails in prison would seem a natural end to Rupescissa's prophetic career, but instead they accompanied a period of great productivity for the writer. While imprisoned, Rupescissa composed a number of treatises that described the apocalyptic crisis he anticipated. Among his major works (others have been lost) are the *Commentum super prophetiam Cyrilli* (Commentary on the Prophecy of Cyril, 1348–49), the *Liber secretorum eventuum* (The Book of Secret Events, 1349), the *Commentum super Veh mundo in centum annis* (Commentary on Woe to the World in One Hundred Years, 1354), the *Liber ostensor quod adesse festinant tempora* (The Book that Shows That the Times Are Soon to Be at Hand, 1356), and the *Vade mecum in tribulatione* (The Handbook in Tribulation, 1356). During the same period, Rupescissa also composed two tracts on alchemy and alchemical medicine, the *Liber lucis* and the *De quinta essentia*.[59]

Rupescissa's imprisonment in Avignon was in some respects advantageous for the dissemination of his prophecy. As the institutional seat of the church, Avignon was one of the great political and intellectual centers of the world. As long as Rupescissa was held in the papal prison, he had access to the company of distinguished visitors, and it is possible that he was permitted to read volumes from the vast papal library. There is evidence that Rupescissa had books in his prison cell, and he was clearly allowed access to pen and parchment.[60] Furthermore, his proximity to the papal court supplied him with news of current political events, which deeply influenced his prophetic exegesis. Despite suspicions about his orthodoxy, Rupescissa appears to have had admirers, even among the high clergy. No one prohibited him from answering multiple requests for prophetic interpretations, and so he provided prophetic advice for Stephen Aldebrandi, the archbishop of Toulouse, in his letter *Vos misistis*; for an unnamed cardinal in another letter, *Reverendissime pater*; and for Peter Pererii, an influential Franciscan, in the *Vade mecum*.[61] In *Liber ostensor*, Rupescissa makes reference to a volume of prophecy sent to him by "a person who loves [him]," indicating that he had some contact with correspondents while in prison.[62] Rupescissa also mentions in the *Vade mecum* visits from Peter Pererii, during which they had many conversations about the misfortunes of the royal house of France.[63] Copies of Rupescissa's writings spread far, apparently with the consent of Rupescissa's jailors. The German Dominican Konrad of Halberstadt, for instance, learned of the imprisoned prophet in 1353 while on business in Avignon, and he was able to carry an excerpted version of Rupescissa's writings home with him.[64] Rupescissa's prophecies were also swiftly translated into vernacular versions. A French translation of the *Vade mecum* was completed as early as 1358, two years after the text's initial composition and certainly more quickly than most modern translations are produced.[65] The traffic of international visitors through the papal palace likely aided the circulation of Rupescissa's manuscripts, as well as the creation of such translations.

Contemporary chroniclers' judgment of Rupescissa was mixed, but at least some document sympathy for the Franciscan and his prophecies. A chronicler and Carmelite friar, Jean de Venette, noted that a "saintly[,] . . . sober, and honest" Rupescissa was held in prison for his predictions, leading many to wonder whether his words were inspired by some "pythonic or evil spirit" or the genuine spirit of prophecy. Venette himself expressed a cautious approval of Rupescissa's *Liber ostensor* and *Vade mecum*, writing that: "though I put no trust in them, yet I have seen many of the events which they prognosticate come to pass. Nor do I think it is impossible that God may have revealed much to him as He has aforetime to the holy fathers, for example, in interpretations of the sacred prophecies."[66] Accounts of Rupescissa's prophecy appear in several other contemporary chronicles, including those of Jean le Bel, Jean Froissart, and Henry of Rebdorf. A favorable description of Rupescissa's gift of prophecy appears in the *Chronicon Moguntinum* under the year 1359, in which the author notes that "in these times a certain Minorite friar was imprisoned by Pope Innocent VI for prophesying many different future happenings in the world, of which the greater part have come true."[67]

Rupescissa's apocalyptic predictions may have resonated all the more deeply with readers due to the crescendo of social and economic hardships that assaulted Western Europe in the fourteenth century. A widespread famine gripped Europe from 1315 through 1322, followed by an epidemic of plague (known as the Black Death) that reached the west in 1347 and 1348, killing at least one-third of the population and transforming the lives and attitudes of the survivors. Meanwhile, France's fortunes in the intermittent wars with England that came to be known as the Hundred Years War were worsening in the years around 1350, when Rupescissa was composing his predictions. A French army under Philip VI (r. 1328–1350) suffered a greatly demoralizing defeat at Crécy in 1346, during which French knights were cut down by a smaller force of English infantry wielding longbows. During a joint invasion by Edward Prince of Wales and Duke Henry of Lancaster, the English army plundered the Loire valley, incensing the French king, John the Good (r. 1350–1364), who initiated a battle at Poitiers in 1356. After another disastrous defeat, the king and his youngest son were captured by the English. Seeking to pay the costs of war, the French nobility extracted heavy taxes from the peasants, whose resources and morale were already low because of the constant fighting and brigandage in rural areas. Ongoing tensions between peasants and lords had led to small uprisings during the early fourteenth century, but burgeoning discontent was palpable in the years after Poitiers and the plague, resulting in the great peasant uprising of 1358, known as the Jacquerie.[68]

In addition, by the mid-fourteenth century, Christians were increasingly disaffected by the Avignon papacy, the so-called Babylonian Captivity, which many perceived to be a deterioration of the spiritual mission and prestige of the Roman church. Since the famous conflict between Pope Boniface VIII and French king Philip IV at Anagni in 1303, subsequent popes cautiously positioned their courts in Avignon rather than Rome, keeping themselves within the sphere of influence (and the good graces) of the French kings.[69] The popes of Avignon, mired in the financial and political debacles of the fourteenth century, acquired a reputation for worldliness and greed. For Rupescissa, the current papal see was the "sinner Avignon," from which the pope and his cardinals would be forced to flee for their lives in the coming crisis.[70]

Rupescissa found himself deeply moved by the tragic events of the fourteenth century and by the apocalyptic prophecy of Joachite prophecy and Spiritual Franciscanism. As the next chapter will demonstrate, Rupescissa's interpretations of prophetic texts framed his apocalyptic predictions, providing a context for the plague, for the wars between England and France, and for the continued opposition to Franciscan poverty in the contemporary church. As he collected and read his sources, Rupescissa engaged with the continuous and flexible system of Christian apocalypticism, finding ways to apply this framework to the events of his day. Along the way, Rupescissa became an eschatological innovator. He minted a new doctrine of the one-thousand-year millennium, and he imagined a new avenue of action for human beings in the clash against Antichrist. Such would be the foundation for the scientific program that he would offer to his readers: a prescription of alchemical remedies crucial to the survival of Christendom in the last days.

Three

JOHN OF RUPESCISSA'S VISION OF THE END

John of Rupescissa's examination before the papal curia provides a clear indication of just what was at stake in the fourteenth-century controversy over poverty in the Franciscan order and the larger church. Rupescissa writes in his *Liber ostensor* that he was brought before a large clerical audience in 1354 and confronted with the inflammatory content of his own writings. Rupescissa's examiner, Cardinal Guillaume Court, addressed him:

"Brother John, in your books which this same cardinal has held in his hand, you prophesy that we must suffer the greatest tribulations, and be humiliated, and lose our wealth and this temporal glory that we have, and that papal power and authority must be returned to certain poor of your order, all of which are insane and ridiculous ideas!"[1]

In the decades before Rupescissa's arrest, Franciscans and Beguins caught up in the Spiritual controversy imagined themselves at the center of the events described in St. John's Apocalypse. They adopted the Joachite portrait of the last days—a time in which poor Franciscans fought evil, preached to the world, and, when necessary, resisted an ecclesiastical elite allied with Antichrist. Like the Spirituals and Beguins before him, Rupescissa took a fateful stand against the clerics at Avignon, likely leading to his denunciation as a *fantasticus* and his continued imprisonment at the behest of the curia.

Rupescissa's debt to Joachim of Fiore and the Spirituals has not gone unnoticed by scholars. The standard work on his prophecy, Jeanne Bignami-Odier's *Études sur Jean de Roquetaillade*, traces Rupescissa's reliance on Joachim and Peter Olivi, as well as pseudonymous Joachite sources, and she provides detailed summaries of each of his extant prophetic works. In particular, Bignami-Odier points to Rupescissa as an example of Joachite prophecy applied to a French political agenda.² Robert E. Lerner has expanded upon Bignami-Odier's biography of Rupescissa, correcting her overemphasis on French Joachism and instead placing Rupescissa within a larger tradition of apocalyptic hermeneutics. Lerner's extensive discussion of the *Liber secretorum eventuum* in the context of Rupescissa's other writings highlights the internal cohesion, novelty, and impact of Rupescissa's prophetic program.³

Rupescissa's apocalyptic theology was indeed novel. He was likely the first writer to assert that human leaders could fight Antichrist and live to enjoy the utopian society of the millennium. He is the only late-medieval prophetic writer (aside from the Taborites in the fifteenth century) to advocate popular violence during the apocalyptic tribulations.⁴ Moreover, he boldly asserted—in defiance of the Church Fathers—that the millennium would endure for a literal one thousand years. Perhaps most important, Rupescissa claimed that human beings could ultimately defeat Antichrist through the study of the natural world, which he described in his writings on alchemy. Each of these arguments made human agency central in both the climax of the Antichrist's arrival and the denouement of the millennium. They also had weighty consequences for Rupescissa's apocalyptic rhetoric: if human beings were to fight the Antichrist, they would need extremely detailed information, including the specific dates of apocalyptic events and the true identities of apocalyptic persons, which Rupescissa was eager to provide. He also increased the significance of the future millennium, describing its character in detail and arguing for its longer duration. He made it clearer than ever just how Christians should prepare and, by describing the millennium in detail, he highlighted the reward for which they fought.

John of Rupescissa's Apocalypse

The *Vade mecum in tribulatione*, the most popular of Rupescissa's prophetic tracts and among the least studied by modern scholars, perhaps best encapsulates the author's interpretive and rhetorical strategies.⁵ It offers a relatively short summary of his apocalyptic program and, because it is among the last of his major prophetic works to be completed, it reflects his fully matured vision of the apocalyptic future.⁶ Composed in 1356 and addressed to Petrus Pererii, a master of medicine and member of the Franciscan order, the *Vade mecum* offers an overview of the future events that would precede and follow an apocalyptic clash with Antichrist.⁷ Rupescissa's description of the last days expresses profound pessimism about the current state of the world and equal optimism about the new society that would form in its wake.⁸ As Rupescissa explains near the beginning of the treatise: "I rejoice in the future restoration, but I am saddened by the imminent, very hard pressure on the whole Christian people, the likes of which have never been since the beginning of the world, nor will they ever be until the end."⁹ To illustrate the dual thrust of this future, Rupescissa juxtaposes terrible descriptions of death and destruction alongside hopeful scenes of the postapocalyptic age of peace.

Rupescissa dated the advent of Antichrist to the period between 1365 and 1370, derived from his computation of the 1,290 days in Daniel 12:11—counting each day as a year from the time of the destruction of the Jewish Temple (which he dates to 75 a.d.).¹⁰ The entire decade leading up to 1370, however, would abound with disasters. Rupescissa recounted with apparent relish the downfall of nearly every stratum of secular and religious society, including the leaders of his own order and church (a bold move given that Rupescissa was still a prisoner of the pope during the composition of this work). White, black, and gray monks, Franciscans and Dominicans, priests and bishops alike would experience unprecedented torture and loss of life from the violence of tyrants, plagues, famines, and various other scourges inspired by God.¹¹ According to Rupescissa, all of the clergy's worldly possessions would be confiscated by tyrants and laypeople, who would also massacre many clerics. Not even the papal curia would be spared from ruin: Rupescissa predicted that the cardinals will be forced to flee Avignon for their lives on the fifteenth of July, 1362.¹² He interprets 1 Peter 4:17 on divine judgment as an indication that "the weight of impending scourges will be turned against the clerics, so much so that in this the entire church will perish—unless it is not able to because the Lord Christ is praying for it—but it will certainly be tossed about so much that many Christians will speculate to themselves that it has finally been choked."¹³

Rupescissa expanded his list of misfortunes to include a popular overthrow of unjust nobles by the people whom they previously oppressed. This revolution is part of a fascinating description of the reversal of roles among predators and prey in the natural world:

For five continuous years there will be horrendous novelties in the world. First, the worms of the earth will take up such strength and boldness that they will cruelly devour nearly all the lions, bears, leopards, and wolves. Larks and blackbirds and owls will tear to pieces the rapacious birds, falcons, and hawks. This is necessary so that the prophecy of Isaiah 33 is completed, proclaiming "Woe to you who are predators, will you not be preyed upon?" . . . Moreover, in those five years the justice of the people will rise up and devour the traitorous noble tyrants in the mouth of twice-sharpened swords. Many of the princes and nobles and powerful men will fall from their dignities and the glory of their riches and there will be affliction of the nobility beyond that which is able to be believed, and those who preyed upon the afflicted populace with their pillagings will be seized.[14]

Rupescissa is remarkably astute about the tensions that would soon lead to the uprising of 1358, the Jacquerie, during which French peasants exacted a violent revenge upon the lords who had oppressed them. Other peasant and worker insurrections across Europe followed in the next few decades, including the Ciompi of Florence in 1378, the English Peasant's Rebellion in 1381, and the Catalan and Hussite uprisings of the fifteenth century.[15] Rupescissa's predictions proved influential during these later periods of revolt: a version of the *Vade mecum* that circulated in late-fourteenth-century Florence includes dates that have been slightly adjusted so that Rupescissa appears to have accurately predicted the Ciompi rebellion. Fifteenth-century redactors in Bohemia similarly revamped or redated Rupescissa's predictions in order to accommodate the current climate of the Hussite uprising, which Rupescissa's writings on peasant violence seemed to legitimate.[16]

In addition, Rupescissa describes the inversion of human power relations in terms of the natural world: the reversal of the nobility and peasantry mirrors an unnatural reversal of predation among animals. In the last days, the order of the world—both natural and human—becomes suddenly dystopic and topsy-turvy. This parallelism is significant because of the connections that Rupescissa draws between nature and human history in his alchemical writings, which I shall address in more detail in the chapters that follow. In short, Rupescissa draws analogical connections between incidents in human history (such as the death and resurrection of Christ) and the behavior of natural materials (such as alchemical agents) in his alchemical works. These analogies use images of human suffering

{35}

and death to clarify the processes and meanings of natural change; in contrast, Rupescissa also cites in the passage above the behavior of animals in nature as a means to understand human behavior. This play upon the parallel structure of nature and humanity sets the stage for Rupescissa's main argument: that the manipulation of the natural world can have real and powerful effects on the outcome of human and salvation history.

Rupescissa allows that the reversal of peasants and lords is "horrendous"; however, it serves an important purpose in the end time. The most irredeemable of the lords will be killed and removed from Christian society. Survivors will lose the comfort of their wealth and position and, humbled, will prepare for life in the third state. But Rupescissa's inversion of the powerful and the weak does not ultimately lead to any changes in societal hierarchy. Although Rupescissa seems here to describe the overthrow of the potentates without much regret, he does not argue for any permanent improvement for peasants in the millennial age that follows. On the contrary, social disorder and inversion appear to be disasters that signal Antichrist's arrival. Rupescissa's vision of the millennium returns to the world both a traditional ruling structure and distinctions of social, religious, and gender status among the ruled, as I discuss below.

The unrest experienced by the clergy and nobility will be accompanied by wars on a large scale. According to Rupescissa, France will face new military conflicts even more serious than those already encountered; Spain and Italy will also fight protracted battles.[17] An infidel army consisting of Turks, Tartars, and Saracens will ravage most of Italy, Germany, Hungary, and Poland. Two Antichrists will appear: an eastern Antichrist, who will mislead the Jews into thinking him the messiah, and a western Antichrist (the great Antichrist), whom Rupescissa refers to as Nero. Nero will become the Holy Roman Emperor and reign for three and one-half years leading up to the year 1370.[18] Through the combined assaults of the infidel army and the western Antichrist, virtually all of the cities of the world will be destroyed. Earthquakes, floods, famines, widespread choking fits, and other bodily illnesses could all be expected in the decade leading up to 1370.

But bloody destruction was to be followed by a more hopeful scenario. God, Rupescissa assures his readers, has made provisions to save the earth. Two Franciscan friars (the two witnesses of Apoc. 11) will arrive to preach and assault the infidel by unleashing the apocalyptic disasters described in that chapter: the closing up of the heavens, the conversion of the waters into blood, and the spread of plagues.[19] Rupescissa makes use of a number of the tropes of apocalyptic discourse, including characters such as the "Angelic Pope" and the "Last World Emperor."[20] The Angelic Pope, whom Rupescissa calls the "Repairer" (*Reparator*) of the church, will be a Franciscan friar. He will expel from the ministry unworthy clerics; he will harshly correct religious who have deviated from pure living; and he will return the right to elect prelates to the episcopal sees.[21] Finally, he will help the king of France, who can be recognized as an archetypal Last World Emperor, to be elected Holy Roman Emperor. After the defeat of Antichrist, the pope and emperor will restore the world, extending the emperor's kingdom into Jerusalem and Asia, healing the schism between the Greek and Latin churches, and uniting the Guelphs and the Ghibellines in Italy. While a number of earlier prophecies proposed the existence of Last World Emperors and Angelic Popes, this is apparently the first time that anyone has suggested that these characters would rise to power during Antichrist's reign and survive into the next age.[22] Rupescissa thus seems to have been the first to claim that individual leaders could struggle against Antichrist and emerge victorious.

THE MILLENNIUM

In his analysis of the duration and character of the millennium, Rupescissa makes some of his greatest innovations. He provides a brief summary of the millennial kingdom in his shorter treatise, the *Vade mecum*, but he devotes a large portion of the lengthy *Liber secretorum eventuum* (chapters 107–27) to the details of its foundation and character. Because Rupescissa's account of the millennium is so much more detailed than that of his predecessors, his can be considered among the first systematic descriptions of the millennium written by a Franciscan.[23] While other prophecies include brief mentions of a future utopian society, Rupescissa's complicated millenarian vision expanded upon Joachim of Fiore's predictions and constituted a culmination of the central Franciscan prophetic tradition.[24] Peter Olivi and Ubertino of Casale offer some description of the millennium, but Rupescissa's *Liber secretorum eventuum* was the most extensive and influential text to address the subject in the fourteenth century. Rupescissa's emphasis on the length and significance of the millennium represents a major development in the Franciscan prophetic tradition, and it points to the centrality of postapocalyptic society to Rupescissa's prophetic program.

Christian theologians had long struggled to determine the length of the period of peace between the advent of Antichrist and the eschaton—the period often referred to as the millennium, regardless of its length. Jerome's calculation, based upon a reading of Daniel 12 (subtracting 1,290 from the 1,335 days), set the time at a brief forty-five days; Jerome also suggested that it would be a time for penance rather than celebration. The period of forty-five (or sometimes forty) days reappears in the works of Bede, Haimo of Auxerre, and Peter Lombard, although it eventually became a time of rest rather than of penance. Even those who argued that the duration of such a period was unknowable agreed for the most part that it would be short.[25] In the late thirteenth century, however, this assumption of brevity was challenged. First, Arnald of Vilanova argued in his *De tempore adventus Antichristi* (originally written in 1288, although not promulgated until 1300) that Daniel's forty-five days should be understood to mean forty-five years, following Arnald's interpretation of Ezekiel 4:6 ("a day for a year I have appointed to thee").[26] Second, in 1298 Peter Olivi advanced a much bolder prediction of the millennium's duration. Following a series of calculations (based variously on the 1,290 days of Daniel 12:11, the 1,260 years of Apocalypse 20:2, and Pope Sylvester's death in 334, plus 1,000 years, minus Christ's 34 years), Olivi subtracted 1,300 from the 2,000 years that he believed made up the sixth age. The millennium now spanned a lengthy seven hundred years, inching ever closer to the literal one-thousand-year millennium. With this innovation, Olivi broke from his greatest influence, Joachim of Fiore, who declined to specify the length of the peace but, like his predecessors, judged it to be brief.[27] But Olivi offset his boldness by offering another method by which he arrived at slightly different dates, along with the admonition that such dating techniques could not be precise.[28]

Rupescissa carried these extensions of the millennium to their most extreme conclusion. He argued that the period between Antichrist and the last judgment would endure a full thousand years, following a literal interpretation of Apocalypse 20:

It is evident what [Scripture] says about the martyrs at the time of Antichrist that "those who adored neither the beast nor his image, etc." will live resurrected in their bodies and reign corporeally with Christ for one thousand years, which will be from the time of the death of Antichrist until the advent of Gog near the end of the world. This resurrection of the body does not include all of the saints and martyrs of other times, but is the exclusive mark and privilege of the martyrs of the time of Antichrist. This is why one finds rightly that "the rest of the dead will not live until the one thousand years pass"—counting from Antichrist's death and the chaining of the dragon until the advent of Gog and the end of time and the world.[29]

Rupescissa relies upon Apocalypse 20 here rather than Daniel 12 to arrive at his figure of one thousand years. He also argues that only those who resist Antichrist during his apocalyptic tribulations will enjoy the millennial peace in their resurrected bodies; all others—including earlier saints and martyrs—will have to wait until the general resurrection for their glory. This claim was an important component of Rupescissa's message to Christians to defy Antichrist. Even if they fell, they would receive the rewards of the millennium—a terrestrial reign of one thousand years—in return for their courage.

Rupescissa continues by defending his position against the opinions of the Church Fathers. He admits that he was stupefied when he received his revelatory insight into the millennium's one-thousand-year duration because it seemed to contradict the teachings of the saints.[30] In time, however, he came to understand that he had been given a divine secret that had never before been comprehended:

One must note that if Augustine had been obliged to give a symbolic explanation of the one thousand years, he had not been given the understanding that the chaining of the dragon has to take place on the day of Antichrist's death, as one sees manifestly in *De civitate Dei*. Moreover, we are not held to follow the explications of all of the doctors in all points; they have advanced them as conjectures rather than as facts asserted against the truth of the text. The obligation holds only if one can prove from sacred text or from some determination of the church that all the secrets of scriptures and of prophecies have been completely revealed to Augustine and the rest of the doctors of past times in such a way that nothing more must be revealed to those who follow. But Gregory, the most esteemed doctor, says the opposite in his homily nine on Ezekiel. He teaches there that people who live later and are closer to the end of the world will understand in a manner more luminous and clear the hidden truths of holy scripture.[31]

Rupescissa's claim that those who lived closer to the end of time would comprehend Scripture better than previous church authorities is crucial. He thought himself to be at a more advantageous position in time than Augustine and, as a result, he felt free to overturn the saint's interpretation regarding the millennium. Rupescissa next dismissed those who believed that the period after Antichrist would endure a mere forty-five days, noting that the saints claimed only that that would be its minimum duration.[32] In a few spirited paragraphs, Rupescissa summarily dispatched the pronouncements of both Augustine and Jerome, which had been the basis for hundreds of years of apocalyptic exegesis.[33] Moreover, Rupescissa referred to this period as the "third state" in the *Vade mecum*, linking his millennial age to Joachim of Fiore's trinitarian division of history.[34]

Rupescissa was no doubt influenced by the calculations of his predecessors, and he incorporated their numbers into subsections within his millennium. Rupescissa predicted that after Antichrist's defeat, wars and other trials would rage for forty-five years (Arnald of Vilanova's number) until the peace was finally instituted.[35] After the peace had endured for seven hundred years (Peter Olivi's number), the sanctity and stability that characterized the millennium would begin to deteriorate: the clergy would slide into pride, gluttony, and corruption, justice would vanish, and the truly saintly would become objects of derision. This period of decline would usher in the final stand of Gog, Magog, and the last Antichrist, which would immediately precede the last judgment.[36]

Along with making the millennium more long-lasting, Rupescissa also made it more desirable, at least from a Franciscan point of view. Despite Rupescissa's care to distance himself from the schismatic members of the Franciscan order, his pronouncements on poverty and corruption within the institutional church are as harsh as those of any Beguin. In fact, Rupescissa recommended the establishment of a radical form of evangelical poverty: the renunciation of temporal wealth not merely by all Franciscans but by all churchmen. Widespread practice of evangelical poverty formed the basis of his spiritual reform program, and it is especially central to his *Vade mecum*. He predicts:

[A]ll the clergy—the highest lords of the universal church, popes, cardinals, princes, primates, archbishops, bishops, and the rest—must be reduced to the mode of living of Christ and the most perfect apostles, since it is otherwise impossible for the church to recover from the aforementioned lost age of the present, as is necessary if the prelates of the church ride in the lands with two or three hundred horsemen just as some today ride, preaching the humility of Christ with knights in such ostentation.[37]

The "manner of living of Christ and the apostles" is—as Rupescissa makes clear throughout his writings—a life of poverty. But universal conversion to poverty will be voluntary, not merely necessitated by the robbery that Rupescissa describes above. After the loss of their worldly belongings, the clergy will receive "understanding" through their afflictions and "humbly recognize their faults and arrange to return to the mode of living of Christ and the apostles."[38]

Rupescissa offers a more detailed account of the fate of the Franciscans, the religious order chiefly charged with the practice of poverty, in his *Liber secretorum eventuum*.[39] Perhaps influenced by the fractured Franciscan order of his own day, Rupescissa predicted that future Franciscans would likewise be split into three factions: the Conventual majority, a second group of true Franciscans who obey the rule to the letter, and a third group of schismatic, hypocritical men who resemble the Michaelists (the followers of Franciscan minister-general Michael of Cesena, who defied the pope and fled to the imperial court of Ludwig of Bavaria in the late 1320s).[40] This tripartite division bears a striking similarity to that described by the Beguins interrogated by Bernard Gui in the 1320s, but in Rupescissa's case a third group of schismatic Fraticelli are instead replaced by the contemporary Michaelists.[41] According to Rupescissa, all but the true Franciscans will ally with Antichrist and his pseudopope, who will be chosen by the cardinals in a schismatic election.[42] The true Franciscans will be forced to flee to caves in the desert along with the true, canonically elected pope. Just as in the condemned Beguins' testimony, only the group of true Franciscans will survive the apocalyptic tribulations. They constitute the future preachers of the last days, and they inherit the obligation to construct a millennial Christian society.

Rupescissa identifies the controversy in the Franciscan order over evangelical poverty not as a symptom of the impending crisis but as its cause. He states: "God said to blessed Francis that if his brothers remained in the observance of the rule shown to blessed Francis just as they had begun, the impending tribulations would not come"; however, because of their transgressions, God has allowed the Dominicans to attack the Franciscans, the order to be shattered into factions, and the apocalyptic tribulations to begin.[43] Rupescissa thus places the blame for the disasters of the apocalyptic crisis squarely upon the Franciscan deviation from the rigors of the rule. This failure, according to Rupescissa, set into motion a divine purge of the order and of Christian society. Rupescissa here espouses a belief not too distant from that of the most radical of the Franciscan Spirituals and their followers, some of whom a generation earlier interpreted the attack on evangelical poverty by John XXII as a trigger for the church's loss of grace and the second fall of humanity. Rupescissa agreed that the order's policies on poverty and the rule would have cataclysmic results, but he did not conclude that the institutional church had lost its sacramental authority, nor did he encourage his readers to rebel. In fact, he assumed in his *Liber secretorum eventuum* the orthodox position on John XXII's *Cum inter nonnullos* and its interpretation of evangelical poverty. Unlike the Michaelists and other schismatic Franciscans, Rupescissa held that *Cum inter nonnullos* was compatible with Nicholas III's *Exiit qui seminat*, a bull that supported Franciscan ideals of poverty.[44] His confidence in the true church, the true order, and the true pope appears not to flag: in the *Vade mecum*, it is a Franciscan pope (even if he is angelic, rather than earthly) who spearheads the reform of Christianity and the battle against Antichrist.

Rupescissa also betrays optimism for the future of the Franciscan rank and file, despite their previous failings. He writes that "through the prayers of the blessed Francis and the holy brothers, the order will be repaired after the tribulation and it will spread out through the whole world, just like the stars in the sky which are so numerous that they cannot be counted."[45] The glory of the future Franciscans will not be in their clothing or buildings or in other "present relaxations" in the rule but through the imitation of the life of Francis and Christ. The extreme evangelical poverty promoted by the Spirituals will therefore become the norm among not only Franciscans but also other religious after the millennial peace has begun.

In addition, strict new laws instituted by the Repairer will result in the disappearance of all vice and violence among the population. The Holy Spirit will descend over all people so that they understand divine truth, prompting them to beat their swords into plowshares and their spears into pruning hooks. According to Rupescissa, the whole church will be covered with the perfections of the life of Christ.[46] Rupescissa also predicts the enlargement of the church due to mass conversion:

[F]rom the beginning of the tribulations and then step by step there will be a gathering of the whole into one flock in the Lord, into one Catholic faith of the general Roman church under the obedience of the one highest Roman

Pope, in such a way that the whole structure of the cosmos (*mundi machina*) will work to convert the Jews, Saracens, Turks, and Tartars to the truth of the general church and to extirpate entirely from the earth all perverse Christians hardened in all mortal sin.[47]

The infidels will convert to Christianity, save a small number who will scatter to the far reaches of the globe. In what has been described as one of the "most concrete philo-Judaic predictions issued by any Christian until this time," Rupescissa claims that the converted Jews will become God's chosen people and will wield power over the entire earth.[48] Although Rupescissa speaks a great deal of conversion, the *Vade mecum* predicts the destruction of the Muslims during the period of tribulations. Rupescissa makes a similar prediction in *Liber ostensor*, which describes the annihilation of the Muslims as a precursor to the final age of renovation and peace.[49] Rupescissa's prediction of the destruction of Islam was not an especially common view in the fourteenth century. The astrologer Jean de Murs offered a similar forecast in his nearly contemporary letter to Clement VI on planetary conjunctions, and because Rupescissa was familiar with the writings of Murs and other contemporary astrologers, he may have been influenced by the same climate of anti-Muslim sentiment.[50] Nevertheless, Rupescissa departs from the opinions of apocalypticists such as Arnald of Vilanova and Ramón Llull, who strongly preferred the conversion of the Muslims to their destruction and who advocated missions by Franciscans familiar with the languages and arguments of non-Christians, a prospect that seems to interest Rupescissa not at all.[51] In chapter 6, I suggest that Rupescissa's reliance on medical and devotional literature led him to advocate the annihilation of certain populations rather than to insist upon their integration into the body of society.[52] His attitude toward Islam appears to be another facet of Rupescissa's predilection for radically removing what he perceived to be diseased extremities from the larger organism of humanity.

Rupescissa claims that all human beings who survive will receive illumination from the Holy Spirit after Antichrist, yet hierarchies of spiritual and social ranking will not vanish from society. A universal emperor and a pontiff (now centered in Jerusalem) will continue to rule over the people. Church leadership will also continue to exist. Rupescissa describes a "college of the church," and he alludes to the convention of seven ecclesiastical councils after the death of Antichrist.[53] Rupescissa notes continuing distinctions between the clergy and the laity, the latter of whom is inferior to the former in terms of purity and sanctity. He writes that while the Holy Spirit will descend over everyone, it will do so more "over the Jews converted to Christ, and even more over the popes, the clergy, and the universal church."[54] Rupescissa predicts that the universal emperor will actually reside outside of Jerusalem "lest the most saintly clergy be polluted by a life of contact with the laity, because it will not be 'as with the people, so with the priest' (Isaiah 24:2)."[55] Apparently, the presence of the Holy Spirit will not raise the spiritual merit of everyone equally, since it will be necessary to insulate the purity of the future clergy from even the highest-ranking laypeople. Although Rupescissa appears to have applauded the violent peasants who, in return for previous abuses, would destroy the riches of the unjust lords during the time of Antichrist, he does not predict for them any spiritual or societal advancement in the millennium. The disorder and inversion of the apocalyptic crisis gives way to a purified societal hierarchy, but not one in which the lower stratum enjoys a greater position.

Rupescissa also notes that familial and marital bonds will continue into the future. Those married people who cannot follow the rule of chastity will be charged with perpetuating the human race through childbearing. Rupescissa adds that these people will live very purely and will generally belong to the third order of the Franciscans.[56] One might wonder why procreation will be necessary during the one-thousand-year reign of the saints, but Rupescissa seems to imagine millennial society as very much like premillennial society. The new age will be populated by kings and popes, clergy and laity, married people and children, but the world will be on the whole more spiritual, more peaceful, and, above all, more Franciscan.

THE PROBLEM OF PROPHECY

In the *Vade mecum*, Rupescissa makes a number of specific predictions for the millennium that do not derive clearly from any passage of Scripture. He calculates specific dates for the flight of the cardinals (an astoundingly specific July 15, 1362), the appearance of Antichrist (c. 1365–1370), and the eschaton (c. 2370). In his other prophetic works, moreover, he identifies apocalyptic characters with actual contemporary persons. In the *Liber secretorum eventuum*, Rupescissa names Louis of Sicily (b. 1338) as the great western Antichrist who will reign for three-and-one-half years and wreak havoc throughout Europe, Asia, and Africa.[57] In ascribing specific dates and names to apocalyptic events and characters, Rupescissa dispensed with the caution of Joachim of Fiore and Peter Olivi, instead following the examples of Spirituals and their sympathizers, such as Ubertino of Casale and Arnald of Vilanova. In fact, and this is essential to his whole program, Rupescissa insisted upon the need for specificity in order to help Christians identify the signs of the end. He argued that through his method "proofs are obtained from holy scripture in confirmation of certain future events through the determination and infallible tabulation of certain ends and more certainly manifest years."[58] The identification of specific events and exact dates is not only possible but crucial. People could certainly better prepare for Antichrist if they knew who he was and when he was coming. But how could one know such things? If these details were revealed in Scripture, they were hardly apparent to a majority of readers. Such a task, according to Rupescissa, fell to those who had insight into Scripture that gave them a special ability to predict with accuracy and a special obligation to do so.

Marjorie Reeves begins *The Influence of Prophecy* by noting:

The medieval concept of prophecy presupposed a divine providence working out its will in history, a set of given clues as to that meaning implanted in history, and a gift of illumination to chosen men called to discern those clues and from them to prophesy to their generation. . . . Divine providence, it was believed, used human agencies and prophecy was often a call to men to involve themselves in the working out of God's purposes in history.[59]

The Joachite apocalyptic, unlike the Sibylline and Pseudo-Methodian prophecies that I discussed in the chapter 2, allowed for the possibility of effective human action in the unfolding of salvation history. Humans were not meant only to watch and wait but also to participate actively in the apocalyptic crisis. John of Rupescissa was convinced of the imminence of a new age and the apocalyptic role of the Franciscans in the initiation of that age. But what could he do about it? How could he play a role in the defeat of Antichrist and the formation of the third state? Clearly he would need answers to a number of practical questions in order to recognize Antichrist and prepare to fight him.

Rupescissa scoured Scripture and virtually any other prophetic source that came into his hands for answers to these questions. During the Middle

Ages, it was considered reasonable that a careful examination of some passage of Scripture or prophecy might yield vital and previously unknown information about the future.[60] Augustine addressed this issue in *On Christian Doctrine*, asking "what could God have more generously and abundantly provided in the divine writings than that the same words might be understood in various ways?"[61] General openness to new interpretations of traditional texts appeared to some exegetes to sanction speculation about the end based on the discovery of previously hidden meaning in the Bible. This premise allowed Rupescissa to claim exegetical innovations in the dates and details of the disasters and the postapocalyptic society.

But there was also a problem with this approach to the Bible. How could the interpreter (or the church, which was responsible for monitoring interpretations) be certain that his or her interpretation of Scripture was correct? Innovative exegesis could occasion fierce rhetorical opposition from theologians and even charges of heresy. In light of this, it was difficult for an exegete such as Rupescissa to offer new and provocative interpretations without risking imprisonment or death. In the *Liber secretorum eventuum* and *Liber ostensor*, Rupescissa attempted to protect himself by describing visionary experiences that afforded him personal knowledge of Antichrist. The first of these visions had shown him the young Antichrist; more visions followed over the next two decades, including several flashes of insight into details surrounding the last days. He also experienced a vision of the Virgin Mary, who appeared to him in front of the temple of Solomon and gave him a book of prophecy.[62]

Rupescissa's claim to visionary authority was certainly not without risk. Prophetic visions were thought to have been common before the incarnation and during the time of the early church, but thereafter miracles and visions increasingly aroused suspicion. Demonic forces, rather than divine illuminations, were assumed to be the source of many ecstasies.[63] Christians who reported visionary experience during the Middle Ages therefore put themselves in great danger of being charged with heresy. Moreover, according to canon law, the burden of proof for orthodoxy fell to the person who claimed direct communication with God.[64]

Rupescissa defended himself from suspicion with what at first seems a paradoxical statement. Although he claims to reveal the events of the end through his predictions, he asserts emphatically that he is not a prophet. He notes the distinction in the *Liber secretorum eventuum*: "I do not call myself a prophet sent by God as was Isaiah or Jeremiah, in such a way that God said to me 'Say this or that to the people,' but I say only that omnipotent God opened my understanding (*intellectum*)."[65] This divine opening of his understanding, Rupescissa explains further, "conferred to most unworthy me the intelligence of the holy spirit of the prophetic scriptures so that I might understand" the apocalyptic meaning of scripture.[66]

This distinction between "prophecy" and "understanding" was vital. Prophecy is generally defined as the expression of an inspired person who claims to speak with the voice of his or her god. The prophet receives a message by means of direct divine revelation or visionary experience, which he or she is compelled to reveal to an audience.[67] Standard examples of prophets in this mold include those of the Old Testament, for instance, Daniel, Isaiah, or Jeremiah. But a number of "prophets" of the later Middle Ages did not consider themselves to be prophets at all according to this definition. Many of them explicitly denied possessing the gift of prophecy, instead claiming "spiritual understanding" (*intellectus spiritualis*), a special insight that allowed them to understand fully the meaning of Scripture.[68]

Rupescissa may be interpreted as one of a long line of thinkers who used what Robert E. Lerner has called the "ecstasy defense" by claiming that potentially heretical innovations in exegesis were the result of divine inspiration.[69] When Joachim of Fiore was asked about the source of his apocalyptic predictions, he explained that "God who once gave the spirit of prophecy to the prophets has given me the spirit of understanding to grasp with great clarity in His Spirit all the mysteries of sacred scripture, just as the prophets who once produced it in God's Spirit understood these mysteries."[70] Joachim's spiritual understanding derived from visionary experiences, the first of which occurred during a session of late-night work on his Apocalypse commentary.[71] Other exegetes reportedly had similar experiences: Ubertino of Casale and Arnald of Vilanova both claimed ecstatic visions as the sources of their apocalyptic ideas, and even Peter Olivi was rumored to have received eschatological insights during a vision in a Paris church.[72] As Lerner has noted, recourse to spiritual insight could protect authors from accusations of heresy and, by the time of Ubertino and Arnald, such claims had become a part of the Joachite (and Franciscan) tradition of apocalyptic exegesis.

A prophetic exegete such as Joachim or Rupescissa differs from a traditional, Old Testament prophet in that he or she interprets a preexisting text (usually Scripture or extrascriptural prophecy) and presents conclusions in his or her own voice; the word of God does not flow directly through the author. Although the traditional definition of prophecy does not include knowledge gained by means of "spiritual intelligence," it is not enough to say that writers such as Joachim and Rupescissa were not prophets—their contemporaries certainly thought that they were. Joachim developed a widespread reputation for prophecy as early as the mid-thirteenth century; as I noted in chapter 2, he was credited with predicting the formation of the mendicant orders, among other important events.[73] Despite suspicions about Rupescissa's orthodoxy, he, too, became known in contemporary circles as an accurate prognosticator. Although neither Joachim nor Rupescissa claimed the title of "prophet" for himself, readers nevertheless applied the label to each. In his *Visions of the End*, Bernard McGinn suggests that in the later Christian tradition, "prophet" came to mean anyone who "foretells the future, or the one who seeks to correct a present situation in the light of an ideal past or glorious future."[74] This definition encompasses the predictive and reformative character of Rupescissa's prophecy; it also accounts for Rupescissa's use of his own voice to make his pronouncements rather than the voice of God (who, as Rupescissa insists, did not order him to "Say this or that to the people").[75]

Both Joachim and Olivi predicted that the final state of history would be characterized by an increasing number of people who could comprehend the meaning of Scripture with increasing clarity through spiritual intelligence. Olivi argued that exegetes of his own time grasped divine truth more fully even than the Church Fathers, who had stood at a less advantageous position in the process of history.[76] Rupescissa's invocation of "intelligence" has some interesting consequences for his "prophecy." When Rupescissa claimed exegetical authority based on progressive spiritual intelligence, he implicitly staked a claim in the Joachite unfolding of salvation history, and he presented his own thoughts as an example of the new understanding that would soon be offered to all humankind. Because those living closer to the end of time understood Scripture more fully and were able to predict more accurately than their forebears, exegetes such as Rupescissa could add significant details to the store of information about Antichrist and the millennium. In fact, following this logic, the newer and less authoritative the writer, the more accurate his interpretations were likely to be. Rupescissa's claim of spiritual intelligence thus provides a powerful rhetorical foundation for his body of prophecy and for his program of naturalist study.

JOHN OF RUPESCISSA'S VISION OF THE END

PROPHECY AND HUMAN AGENCY

In the pre-Antichrist millennialism of the Sibylline and Pseudo-Methodian prophecies, human leaders capitulated to Antichrist and awaited divine aid from Christ. Post-Antichrist millennialism, popularized by Joachim of Fiore, overthrew the assumption that human beings were impotent against Antichrist and raised the possibility of an earthly paradise in the future. John of Rupescissa further developed the themes of the post-Antichrist society and the role of humankind in establishing it. His account of the apocalyptic battle argued that while divine intervention was ultimately necessary, human beings had a great capacity not only to resist Antichrist by refusing to succumb to his temptations but also to inflict harm upon him and his supporters. Rupescissa, for instance, predicted that the French princes—joined by representatives from the church—would organize a relatively successful rebellion against Antichrist:

[T]he ecclesiastical primates and pastors will literally assemble with all of the crowd of the elect and the vigorous Christian knights. Certain of these knights who are issued from the race of the French princes, symbolized by the strongest Maccabees, will rise up. Although they will already have been bruised and chased to the desert, although they will be few in number and weakly aided, they will nevertheless wreak vengeance on the evil nations and stir up violent tempests against Antichrist as if he were a second Antiochus.[77]

Rupescissa emphasized the contingency of salvation history: although he claimed to predict predestined events, he also allowed for human will to affect the course of those events. In *Reverendissime pater*, Rupescissa outlined what Elizabeth Casteen has described as an "unfixed future" contingent upon readers' reactions to his dire predictions.[78] Similarly, in the *Liber secretorum eventuum*, Rupescissa describes the conduct of the French king during Antichrist's attack on Christendom:

In these days, the French king will choose to do one of two things: drop his head before Antichrist (let us hope not!) or be deprived of his kingdom by Antichrist for a time. Dejected, he will patiently suffer the trial of Antichrist with the Holy Roman church.... It is more honorable for the king to be overcome by the beast and to suffer in the company of the church for a brief period than to consent to error. It is therefore necessary for the king to be well informed about these secrets and the future scandals of the antipope since darts that are anticipated are less injurious and traps that are foreseen are easier to avoid.[79]

This passage points to a number of the central ideas in Rupescissa's prophecy. First, although certain future events are inevitable, human beings will nevertheless be required to choose their fates in the last days. Even the decisions made by an apocalyptic character as important as the French king (the potential Last World Emperor) are still uncertain. Second, the contingency of salvation history makes access to apocalyptic exegesis all the more important. Rupescissa supplied the church with advance notice of future events in the hopes that he might influence their outcomes. As Rupescissa argues in the passage, "traps that are foreseen are easier to avoid," if humans are made aware of future events, they will be more likely to make correct decisions at critical times. This makes Rupescissa's task essential.

The passage also reflects the two primary views of human agency in apocalyptic literature: first, the Sibylline view that human beings cannot face Antichrist and, second, the Joachite view that humans can act by recognizing and resisting evil. Rupescissa cited both Sibylline and Joachite prophecies as textual sources for his own treatises.[80] But he clearly adopted the view of the latter tradition, which gave him great latitude to suggest a number of "remedies" that people could use against Antichrist and his temptations. Rupescissa encouraged readers to pay attention to such predictions because, as he claimed, those who are already informed will "resist through these vigorous labors, and they will do this most of all when they know the time and the certain number of years, and when they know that the tribulation of the error of Nero and Antichrist will not last more than three and one half years and that afterwards the whole church must be repaired."[81] Christians who knew what to expect would be less likely to despair or be seduced into heresy by Antichrist, especially if they knew that a millennial society loomed in the near future. Moreover, a number of practical preparations were possible for those who had early warning. Rupescissa recommended that Christians construct safe houses in mountain caves so that they might wait out the worst of the tribulations in seclusion. He advised readers to stock the caves with "provisions of beans, legumes, honey, salted meats, and dried fruits"—a sort of bomb shelter for the period of Antichrist.[82] Rupescissa noted that those who remained in the caves until 1370 could emerge having safely avoided the tribulations.

Information about the end culled from Scripture and prophecy could thus make an apocalyptic crisis more controllable and human defeat less inevitable. The new emphasis on the character of the postapocalyptic society also highlighted the proximity and desirability of the future.

For Rupescissa, the recent disasters of the fourteenth century, as well as the apocalyptic disasters of the near future, elicited not hopelessness but a call to action. Rupescissa's innovative claim that humans could actively participate and intervene in the happenings of last days and the battle against Antichrist was extremely important. It designated a role for humans in some of the most significant events in Christian history, and it articulated a discourse of optimism in the context of a predominantly disastrous age. Rupescissa's message about the potential of human action was attractive to readers, who edited, amended, and disseminated copies of his prophecy and who advocated his recommendations concerning reform, as well as popular justice and revolt.

Rupescissa's final remedy for the survival of humankind and the reform of society arose not from the texts of Scripture and prophecy but from technologies derived from the study of natural world. In his two texts focused on alchemical medicine, the *Liber lucis* and *De quinta essentia*, Rupescissa offered to his readers perhaps their most powerful tool against the apocalyptic crisis: the knowledge to heal bodies and, as a result, to fight Antichrist and survive into the peace of the coming third state.

ALCHEMY IN THEORY AND PRACTICE

Held for decades in prisons, John of Rupescissa enjoyed few comforts from the outside world. He valued the visions he received from God, as well as the rare generosity of other people, one of whom helped Rupescissa through—of all things—alchemy. Rupescissa explained:

And since I was held unjustly in this dark cell by my enemies, and since my body was corrupted by the evils of the chains and the cell, through the kindness of a servant, I was able to have *aqua ardens* from a certain man of God and my friend; and with only a smearing of it around the place [of the wound], I was cured.[1]

We cannot know if this was Rupescissa's first encounter with *aqua ardens* (that is, alcohol distilled through alchemical means), but its beneficial effects clearly made a profound impression upon him in prison. Rupescissa became preoccupied with alchemy at least by the early 1350s, when he dedicated his first alchemical manual, the *Liber lucis*, to the problem of metal transmutation. He entered the world of medical alchemy shortly thereafter with his *De quinta essentia*, a text that urged readers to apply naturalist knowledge to the problems of the apocalyptic church.

We have already seen how Rupescissa's reading of history based on Joachite texts led him to expect the arrival of Antichrist and a series of apocalyptic disasters in the near future. But Rupescissa's analytic resources were not confined to prophetic scripture. His apocalyptic program also depended heavily upon texts that theorized the natural world. Having been cured of his personal injury by *aqua ardens* in prison, Rupescissa looked to alchemy for a cure to the larger woes of the wider world. As the next chapters will show, Rupescissa found in this discipline highly practical tools that could be used to manipulate nature and, consequently, to participate in the apocalyptic crisis. This chapter will examine Rupescissa's idea of the "quintessence," an alchemical innovation that had enormous import for the development of alchemy and medicine. Scholars routinely cite Rupescissa as a pivotal figure in the growth of the pharmacological alchemy in the early modern period, but no historian has devoted sufficient attention to the quintessence and its textual sources to appreciate its revolutionary nature. I shall demonstrate here that the quintessence was formed from intellectual ingredients already present within medieval alchemy and medicine, which Rupescissa integrated and systematized in such a way that his conclusions stood in striking contrast to the received wisdom of medieval naturalism. In the midst of the disasters of the late Middle Ages, Rupescissa imagined the perfection of heaven as hardly remote and inaccessible. Instead, it was within human reach.

The thirteenth and fourteenth centuries were a fertile period for the institutionalization and organization of knowledge, as well as for the development of specific branches of naturalist thought. During this period, natural philosophy and medicine became disciplines at the newly established universities of Europe, laying the groundwork for a systematic approach to the natural world. Universities offered training in the liberal arts, as well as specialized curricula in medicine, law, and theology. While arts curricula varied according to institution, students generally learned elements of logic and disputation, as well as the basics of natural philosophy, arithmetic, geometry, astronomy/astrology, and music theory. Aristotelian natural philosophical texts, which had become newly available between the late eleventh and early thirteenth centuries, were at the center of this curriculum. Statutes issued at the University of Paris in 1366 show that students were required to study the *Physics, On Generation and Corruption, Heaven and Universe, Parva naturalia* (that is, *On Sense and Sensation, On Sleep and Sleeplessness, On Memory and Recollection,* and *On the Length and Brevity of Life*), and *Metaphysics*.[2] Similar documents issued at Bologna in 1405 ordered three years of training each in natural philosophy and astronomy/astrology at the University of Arts and Medicine.[3] In addition, the medical curriculum at Bologna offered four years of medical theory (the philosophy of medicine and principles of the body and disease) and practice (the study of specific techniques of diagnosis and treatment). Bolognese statutes indicate that medical students studied the works of the *articella* (sometimes called the *ars medicine*), a collection of standard medical texts, including the *Aphorisms* and *Prognostics* of Hippocrates and the *Tegni* of Galen, among other works, that was first compiled in the twelfth century and was periodically augmented. Students assimilated their lessons by reading treatises and commentaries and by listening to oral lectures and disputations.[4]

But the struggle to comprehend and manipulate nature was not limited to the universities, as shown by the continuing interest in alchemy in the late Middle Ages. Unlike medicine, natural philosophy, and astronomy/astrology, alchemy never gained an official place among university curricula, even if many of its enthusiasts were university trained.[5] Nevertheless, it drew many of its principles from the university-taught disciplines, with which it shared basic assumptions about therapeutics and the structure of the natural world.

Alchemy is usually defined today as the attempt to transmute base metals into gold by means of a chemical agent, often called a "philosophers' stone." But medieval alchemy comprised a wide variety of goals, techniques, and theories that overlapped with those of other fields, including medicine and natural philosophy. Alchemy was primarily concerned with three areas: first, the transmutation of precious metals; second, the prolongation of human life and health; and third, the general study of materials composed of the four elements.[6] There are competing theories about when and where alchemy originated, but references to alchemy appear among law codes composed in China as early as the fourth century b.c.e.[7] Chinese alchemy directed its resources toward the extension of human life to great lengths—even to eternity—through potable gold and other age-defying remedies. The project of alchemy in the West, which emerged somewhat later and chiefly in Hellenistic Alexandria, was more concerned with the production of precious materials than improvements to the human body. Among the earliest surviving descriptions of Greco-Egyptian alchemy are the crafts recipes collected in two papyri known as *Leidensis* and *Holmiensis* (denoting their current locations in Leiden and Stockholm); they date from about 300 c.e. but are thought to be based upon older models.[8] Such recipes include instructions for the synthesis of various gems and metals, including artificial pearls, emeralds, sapphires, gold, and silver—the latter two metals would emerge as the ultimate goals of transmutatory alchemy in later centuries.

Arabic naturalist texts, including the corpus attributed to the eighth-century author Jābir ibn Hayyān (actually composed by various anonymous

Ismāʿīlī scholars of the ninth and tenth centuries) furnished new foundations for alchemy, including the transmutatory "elixir"—a chemical agent (also known as the philosophers' stone) that purportedly transformed any metal into gold by balancing its constituent elements. The Jābirian corpus was also responsible for introducing new working materials into the discipline, including chemicals such as sal ammoniac and saltpeter. The *Tabula smaragdina* (*Emerald Table*), attributed to "Hermes Trismegistus," and possibly authored by the eighth-century Arab Balīnūs, constituted perhaps the single most significant text in early medieval alchemy, despite its brevity and near impenetrability. Its poetic language seemed to many readers to refer to the refinement of matter through distillation, and it ultimately gained the authority of a classic work of alchemy. The genuine texts of the Muslim naturalists Rhazes (Abū Bakr ibn Zakarīyā al-Rāzī, 865–925) and Avicenna (al-Husain b. ʿAbdallāh Ibn Sīnā, 980–1037) also made important contributions to the elaboration of medical and alchemical theory, as did spurious works attributed to the two authors, including the Pseudo-Avicennian *De anima in arte alchimiae*, which influenced a number of high-medieval Latin naturalists, including Albert the Great.[9]

Alchemy was less developed in the Latin West during the early Middle Ages because of the linguistic and cultural rift between the eastern and western regions of the Roman Empire during late antiquity. Such geographical and cultural distance made the formative texts of Greek alchemy for the most part inaccessible to Latin-speaking authors. Western glass makers, engravers, metallurgists, dyers, and other craftspeople, however, preserved Greco-Roman artisanal technology that was to be applied to metal transmutation and alcohol distillation in later centuries. Artisanal recipes appear in several medieval compilations of the technical arts, including the eighth- or early-ninth-century *Compositiones variae*, the ninth-century *Mappae clavicula*, and the early-twelfth-century *De diversis artibus*, but it is not until the high Middle Ages that a vital tradition of alchemy emerged in the Latin West.[10]

Alchemy was reinvigorated in Western Europe in the mid-twelfth century through the introduction of Latin translations of classical and Muslim naturalist and philosophical texts. Alchemical writings were among the translated works, and their ideas depended upon the newly rediscovered Greek and Arabic bodies of knowledge. For instance, alchemy followed Aristotle's assertion that all matter is composed of four elements (earth, air, water, and fire), which are in turn composed of "prime matter," the basic substance of the material world. Among the most important Aristotelian works for the development of alchemical theory were *On Generation and Corruption*, which treated different kinds of natural change, and *Meteorology*, which discussed the generation of minerals by dual "exhalations" in the interior of the earth, and which included an interpolation (now known to be Avicenna's *De congelatione et conglutatione lapidum*, part of the *Kitāb al-Shifā'*) that stirred medieval debate about the possibility of transmutation.[11]

In the mid-thirteenth century, the eminent scholar and Aristotelian commentator Albert the Great (c. 1200–1280) composed two texts, the *Meteora* and *De mineralibus*, which combined ideas from Arab alchemical texts with Aristotle's theory of mineral generation. *De mineralibus* drew upon Pseudo-Avicennian texts, such as *De anima in arte alchimiae*, to supplement Aristotle's scant treatment of metal formation. The resulting summa incorporated alchemical concepts into Aristotelian natural philosophy and signaled a major development in the theory of transmutation.[12] Other milestone works of the thirteenth and fourteenth centuries included Pseudo-Geber's *Summa perfectionis* (actually by Paul of Tarento; late thirteenth or early fourteenth century) and Petrus Bonus of Ferrara's *Pretiosa margarita novella* (c. 1330), both of which discussed transmutation in terms of medicine or natural philosophy and attempted to integrate alchemy into the domain of scholastic naturalist inquiry (an endeavor that was ultimately unsuccessful).

As alchemy developed in Europe in the late Middle Ages, it became a science once again concerned with not only the transformation of matter but also the preservation of human health and life. The idea of the "elixir" that could purify metals was extended to encompass human bodies: by the thirteenth century, Roger Bacon promoted the idea of a universal alchemical medicine that could cure disease and prolong life. He also divided alchemy into two subdisciplines: *alkimia speculativa* studied change among all things made from the four elements, while *alkimia operativa* applied this knowledge to the creation of metals and other tangibles. Although alchemical elixirs and the prolongation of life were important to Chinese alchemy, Greek and Arabic alchemical texts made few connections between the elixir and the human body. Scholars now credit Roger Bacon with promoting an alchemy of health in the West, a tradition that proved formative for John of Rupescissa, who borrowed from alchemical therapeutics as well as from traditional transmutatory alchemy.[13]

Rupescissa also integrated sources of knowledge from within the university (that is, medicine, natural philosophy, and astronomy/astrology) and from without (that is, the alchemical literature at the fringes of learned society). In his writings, transmutation, cosmology, and therapeutics all coexisted, subjugated to the service of one transcendent goal—the end of the age and the revelation of the meaning of human history. The history of John of Rupescissa is thus an intersection of many histories, providing us with a vantage point from which to view a much larger picture of intellectual, religious, and cultural currents in the late Middle Ages.

APOCALYPTIC ALCHEMY:
THE *Liber lucis* AND *De quinta essentia*

During his long imprisonment in Avignon, John of Rupescissa composed a number of treatises concerned with apocalyptic prophecy and the coming of Antichrist. He also directed his attention toward naturalist studies, producing two texts of what I call "apocalyptic alchemy," the *Liber lucis* and *De quinta essentia*. Rupescissa described in these texts two different alchemical "elixirs": one made from inorganic materials for the purpose of metal transmutation, and another made from both organic and inorganic materials for the purpose of medicine.[14]

Rupescissa's alchemical writings are not entirely distinct from his prophetic works. Rupescissa begins the *Liber lucis* by warning of the coming Antichrist and asserting the utility of alchemy for opposing apocalyptic disasters. He predicts an array of misfortunes for Christians that are nearly identical to those described in his prophecy:

I have considered the tribulations of the elect prophesied by Christ in the Holy Gospel, most of all the tribulations before the time of Antichrist, impending in the years in which the union under the Holy Roman church must no doubt be most afflicted and must flee to the mountains and must certainly be plundered of all temporal riches by tyrants in a short time. The little ship of Peter may be rocked by the rivers, but it will nevertheless be freed in the final days of tribulation and remain the general master. Wherefore, to solve the serious need and future poverty of the holy people and elect of God to whom it is given to know the mystery of the truth, without obfuscation, most of all the red and white Philosophers' stone, I wish to say that I will reveal them concisely and clearly at this time.[15]

In this passage, Rupescissa cites as the impetus for his alchemy the same apocalypticism that fueled his prophetic writings. He describes the seizure of the church's goods and the flight of the true Christians to the mountains, themes that he already treated at length in his prophecy. This link provides a logical imperative to study the natural world: if the church's possessions were to be confiscated by tyrants, as Rupescissa predicted, Christians would be vulnerable to attack. Alchemical technology could produce precious metals, which could be used to purchase necessities and presumably fund the war against Antichrist.[16]

Having suggested that the "philosophers' stone" will be a crucial technology for the Christian elect, Rupescissa goes on to explain what this mysterious substance is. The philosophers' stone (which he also calls the "elixir" and, echoing Albert the Great's *On Minerals* and Pseudo-Geber's *Sum of Perfection*, the "perfect medicine" for metals) automatically converts base metals into pure silver and gold:

Anyone who wishes to create silver should put the prior medicine [the philosophers' stone] onto silver and then pour out the silver as medicine over a body, that is, onto copper or iron, or over other bodies [of base metals] to convert them to true silver. To create true gold, first pour the medicine over the most pure true gold to make medicinal gold, and then put the medicine on copper and other bodies to convert those bodies into true gold.[17]

An aspiring alchemist should apply the philosophers' stone to preexisting gold and silver, which allows the "medicine" to multiply the precious metals by transforming base metals into silver and gold. In order to make this process clear, Rupescissa divides the operation into seven stages (the eighth and final chapter of the *Liber lucis* details the construction of the alchemical oven). He begins with the traditional Jābirian view that the philosophers' stone comprises mercury and sulfur, although he adds that the sulfur is not ordinary sulfur but the "sulfur of the philosophers," which is "Roman vitriol." The initial stage is the sublimation of saltpeter, Roman vitriol, and mercury; this results in a mercury sublimate that is as "white as snow." The next stage involves distilling the "quintessence" from the sublimated mercury, which leaves a blackened "terrestrial" waste in the bottom of the vase.¹⁸ Later stages include further calcinations, dissolutions, and distillations, until one creates either a "white" philosophers' stone to make silver or a "red" philosophers' stone to make gold.

At first glance, the *Liber lucis*, which cites the authorities of "Hermes" and "Geber," and which uses the standard mercury-sulfur theory to produce mineral acids, is a fairly typical alchemical text of the fourteenth century.¹⁹ A similar refinement of matter through multiple dissolutions and distillations also appears in the *Rosarius philosophorum* attributed to Arnald of Vilanova and the *Testamentum* attributed to Ramón Llull. Yet the apocalyptic imagery contained with the *Liber lucis* is hardly typical. Frequent apocalyptic references within the text link transmutatory alchemy to Rupescissa's prophetic writings and to the medical alchemy of *De quinta essentia*. For instance, Rupescissa concludes his instructions for creating the philosophers' stone with this message:

Believe me, poor evangelical man, no man before me has written this truth so openly. Know and be attentive to the fact that many masters, to whom this mystery has been revealed through revelations, invoked harm and cursed with horrible maledictions, fearing that these secrets might come into the hands of the unworthy. But I certainly do not pay attention to these maledictions because I do not reveal the secret to wicked men or to the sons of men but to the Sacred mystical body of Jesus Christ, that is, the Roman church, which does not have the stain of mortal sin, nor does it wear the foreign color of heresy or errors. For I have revealed this only for the Saints as remedies for tribulations in the coming times of Antichrist.²⁰

Rupescissa indicates here that his intended audience is the "poor evangelical man," (*vir pauper evangelice*); this is likely a reference to those Franciscans who practice "evangelical poverty" and who appear so prominently in his prophetic works.²¹ These men are equivalent to the "spiritual men" of Joachite and Spiritual Franciscan prophecy, men who would fight Antichrist and institute the period of millennial peace. Moreover, Rupescissa explains that the process of alchemical transmutation of base metals has been kept secret up until now, and he alludes to the secrecy clauses that are omnipresent in medieval alchemical works.²² But because of the urgency of the church's situation in the "coming times of Antichrist," Rupescissa dares to reveal the secret process of transmutation. The logic bears a strong similarity to that behind Rupescissa's revelation of specific names and dates of apocalyptic events in his prophetic writings. In both cases, the gravity of the coming danger necessitates the disclosure of information traditionally kept secret. We may therefore see a clear line leading from Rupescissa's millennial prophecy in the *Liber secretorum eventuum* and *Vade mecum* to his descriptions of alchemical technology in the *Liber lucis*.

The apocalyptic references in the *Liber lucis* are distinctive, but its alchemical operations were not unusual. One cannot say the same about Rupescissa's other alchemical work, *De quinta essentia*. Robert Multhauf views Rupescissa's systematic approach to the medicinal powers of plants and minerals in *De quinta essentia* as unique among medieval alchemical writings.[23] With it, Rupescissa introduced a potent idea into the discipline: the concept of "quintessences" as a class of chemicals, which I shall discuss below. Moreover, he continued the trend of including mineral-based substances (especially those made from gold, antimony, vitriol [metallic sulfate], and mercury) among his remedies, a project already proposed by thirteenth-century medical practitioners but expanded by Rupescissa through the use of new techniques and new applications to internal medicine.[24] Both the idea of the quintessence and the use of alchemically produced materials in internal medicine were adopted by later and more influential works of "iatrochemistry" (literally, "medical chemistry"), especially in the sixteenth century.[25] Paracelsus is generally credited by historians with incorporating alchemical techniques and concepts into medicine in the early modern world, thus transforming the alchemical tradition. Multhauf notes that Rupescissa's text seems to belong to this later phase of pharmacological alchemy, and he speculates that Rupescissa may be the originator of this extremely important and long-lived genre.[26] Although Multhauf's insight is valuable, it is not enough to note Rupescissa's place in the development of early modern medical alchemy without giving full consideration to his quintessence. For the remainder of this chapter and the next, I shall provide such an analysis, and I shall demonstrate the relationship of Rupescissa's ideas to the naturalist conventions of the mid-fourteenth century. Only then will it be possible to determine to what extent Rupescissa's work constituted a break with traditional medieval therapeutics.

If *De quinta essentia* looks ahead to a new tradition of medical alchemy, it also looks behind to a continuing tradition of medieval apocalyptic prophecy. In its introduction, Rupescissa asserts the utility of alchemy for evangelical men in an argument much like that of the *Liber lucis*. In *De quinta essentia*, however, alchemy is aimed not at the financial strength of the collective church but at the physical strength of the individual Christian. Rupescissa begins by expressing regret for the many years that he spent studying "mundane philosophy" at the University of Toulouse. During this period, he wasted countless hours on "the clattering of inane words, the conflict of useless disputations, and the vain praise and foolish glory of readings."[27] He feared that such use of his time placed him at a disadvantage in the more important quest for eternal life; as a result, he decided to set forth his revelatory insights into nature:

I saw the useful things that were in Philosophy, as the divine spirit showed, even from the time of my youth. At this time God came to me with pity and mercy so that I would reveal [these things] to the poor of Christ and the Evangelical men, so that those who have contempt for riches because of the Gospel know easily and quickly through divine generosity, without human teaching and without much expense, how to heal their bodily miseries and human infirmities, to flee obstacles marvelously with holy prayer and meditation, and to resist even the temptations of demons, which attack because of certain infirmities. This is so that in all things and through all things they are able expeditiously and most devotedly to serve with all men our Lord Jesus Christ. And because I make this work only for the Saints, and because, with God willing, they receive from this work the many good things contained within it, then clearly I will be in their good works, prayers, and merits which they do more with the help of this book.[28]

De quinta essentia is directed to the "evangelical men" (*viri evangelici*), the same audience as that of the *Liber lucis*. Rupescissa also makes reference in the next paragraph to his literal millennialism, noting that there are "undoubtedly more than one thousand years to come before the end of time."[29] From the beginning of the text, millennial apocalypticism and the plight of the spiritual men are central components of Rupescissa's alchemical project.

De quinta essentia focuses on the restoration and prolongation of health through alchemical medicine, which Rupescissa suggests is all the more important in the context of the last days. Healthier and longer-lived evangelical preachers would presumably be more formidable allies of the church and adversaries of Antichrist. *De quinta essentia* therefore includes two parts: book 1 focuses on the theoretical underpinnings of alchemy; book 2 outlines the practical steps of healing. This organization reflects the common division of scholastic disciplines into two areas, the *theorica* (knowledge of underlying causes and principles) and the *practica* (this knowledge directed toward practical goals). Rupescissa's *practica* in book 2 offers a list of medical conditions, including skin diseases, infected wounds, lice, leprosy, poisoning, various fevers, and plague, followed by prescriptions for treatment, some of which are culled from another book of remedies, *De virtutibus simplicium medicinarum*, also known as *Cogitanti mihi*, which has been edited among the works of Isaac Israeli.[30] Several of Rupescissa's remedies are clearly aimed at the evangelical men; for instance, he offers a recipe for a salve to heal feet damaged by excessive walking and itinerant preaching.[31] He also addresses directly the dangers of the apocalyptic era, for instance, by including instructions to conceal gold through amalgamation, which Rupescissa writes is valuable "in all times of war and tribulation and most of all in the time of Antichrist."[32] He also prescribes for soldiers portable jars of *aqua ardens* to treat war wounds. He perhaps has in mind the great apocalyptic battles predicted in the *Liber secretorum eventuum* and the *Vade mecum*.[33]

Rupescissa's prophetic and alchemical writings share not only the goal of countering apocalyptic disaster but also assumptions about the source of knowledge needed to achieve that goal. Rupescissa indicates in the prologue to *De quinta essentia* that God revealed to him natural secrets in order to safeguard the health of the church. He similarly claimed in his *Liber secretorum eventuum* and *Vade mecum* that he received divine revelation that "opened his understanding" of the apocalyptic meaning of Scripture. As we have seen, Rupescissa used this statement to align himself with other apocalyptic exegetes (including Joachim of Fiore and Arnald of Vilanova) whose "spiritual understanding" granted them special insight. I argue that Rupescissa's ability to understand alchemical secrets through divine inspiration mimics his ability to understand the prophetic meaning of Scripture. He claimed that all knowledge of true medicine came from revelation; contemporary physicians, who "burn with the desire for money and honors," were completely ignorant of medical cures because God did not reward the greedy with revelatory insight.[34]

While Rupescissa regarded his quintessence alchemy as the product of divine illumination, his writing also made it clear that its processes were comprehensible in terms of contemporary natural philosophical and medical theory. For Rupescissa, prophetic knowledge was in no way incompatible with knowledge based upon more mundane sources, such as the writings of past authorities, logical deduction, and experience. Rupescissa himself seems to have been primarily an armchair physician: there is no evidence that he ever practiced medicine or alchemy himself, apart from his self-treatment of a wound with *aqua ardens* during his long imprisonment.[35] Perhaps because of his limited ability to engage in the art firsthand, Rupescissa favored theoretical elaborations of alchemy based on reason and on the writings of alchemical authorities, rather than ones based upon his own experience. Such concern with theoretical explanation places Rupescissa squarely in opposition to the practical (or what we might now call empirical, although this word had a decidedly negative connotation in the fourteenth century) tradition of medicine preferred by many practicing physicians, who were often more concerned with immediate therapeutic results than with general theories of causation.[36] Arnald of Vilanova, whose medical interests were deeply practical, explained in his *De intentione medicorum* that the theories of physicians were occasionally less "true" than those of the philosophers but, because they had superior therapeutic value, they were more useful.[37]

In contrast, Rupescissa's chief concern was to determine philosophical truth and to reveal it to his readers. It mattered little that he never practiced alchemy because he arrived at his conclusions through visionary insight. His writings were no mere lists of medicinal recipes; they were concordances of science and revelation. Echoing criticisms of those who ignore natural causes from William of Conches and Thomas Aquinas, Rupescissa ridiculed the physician John Mesüe, who (he claimed) discouraged readers from seeking the causes of certain medicinal cures.[38] In contrast, Rupescissa asserted his own ability to understand "the cause of marvelous things through the incomprehensible light of God," and he characterized his writings as an exposition of materials needed to augment the medicinal power of the quintessence.[39] Rupescissa's attention to both prophecy and natural philosophical reasoning as sources of alchemical knowledge is significant. He described alchemy in the language of miracles, but he also asserted its obedience to the secondary causes of humoral balance and celestial influence (more on this in a moment). For Rupescissa, any understanding of nature or Scripture derived from divine inspiration, and both were aimed at the same eschatological goal. Rupescissa's epistemology synthesizes the divine and the natural, asserting their equivalence. As a result, modern understandings of science and religion as distinct categories are limited in their ability to describe Rupescissa's approach.

Rupescissa was deeply interested in transmuting base metals into silver and gold, but he never called himself an alchemist and never mentioned alchemy except in a pejorative manner. He preferred the term "philosophy" for his endeavor. His distaste for the term "alchemy" may reflect the bias against alchemy in the universities of medieval Europe. It is also possible that Rupescissa read the disdainful characterizations of alchemy by Thomas Aquinas and Giles of Rome, who argued that the art produced deficient or corrosive products.[40] But it seems more likely that alchemy was for Rupescissa too limited a term for the grand project that he imagined: the radical transformation of the human body and earthly society. The world of fourteenth-century alchemy, like Rupescissa's personal world of apocalyptic prophecy, was one of indefinite borders and multiple approaches to understanding and controlling the natural world. It was therefore possible to transmute gold and create life-prolonging alchemical medicines while still viewing oneself chiefly as a "philosopher." Nevertheless, Rupescissa's readers certainly considered him to be an alchemist, and his theories had significant consequences for the development of alchemy, even if they did not bring about the apocalyptic transformations Rupescissa had in mind.

THE IDEA OF THE QUINTESSENCE

The crown of Rupescissa's alchemical thought is his "quintessence," a marvelous alchemical remedy with seemingly preternatural powers. The quintessence is a complicated idea and a point of convergence for many intellectual traditions. As a result, I have chosen to present it in a somewhat unusual manner. First, I shall describe the quintessence and its supposed functions, then I shall demonstrate its reliance upon Aristotelian natural philosophy and Galenic medicine. Finally, in the next chapter, I shall trace some of the major textual sources that Rupescissa may have used to generate his ideas—those authored by or attributed to Roger Bacon, Arnald of Vilanova, and Ramón Llull. While my presentation inverts chronology, I believe that such a strategy is necessary to demonstrate just how unusual Rupescissa's conclusions were in the context of his predecessors.

In *De quinta essentia*, Rupescissa describes the quintessence as a medicine that counteracts virtually all of the effects of human disease and aging. He notes that there are three ways in which humans can die:

The first way is by natural death at the end of life determined for us by God, which we are not able to escape through any natural genius. Another way is by violent death; and in these two ways medicine is in vain. The other way is by chance or as occasions arise death arrives sooner than the end predetermined by God. This is because of too much regeneration and dissolution, or through too much austere abstinence or despair, or through negligence in avoiding danger of death, one kills oneself.[41]

This third cause, which results from the neglect of health and the failure to avoid danger, is the focus of Rupescissa's alchemy. The passage alludes to the contemporary medical theory that humans should observe a "regimen" of health through the regulation of the "non-naturals" in order to extend life and avoid the worst effects of aging. The nonnaturals were six factors considered necessary to life and health, including air quality, food and drink, evacuation and retention, wakefulness and sleep, rest and exercise, and psychological well-being.[42] Those who failed to control the nonnaturals through an appropriate regimen—sometimes called "hygiene"—could attempt too much "austere abstinence" or "regeneration and dissolution" and, as Rupescissa says, inadvertently kill themselves.

But Rupescissa has in mind here something beyond the standard regimen of health known to ordinary physicians. He claims that he has secret knowledge of a substance that will cure all illness and prolong youth indefinitely, or at least until the time of death preordained by God:

[T]hrough the virtue which God contributed to ornamented nature and subjected to human magisterium, man is able to cure the inconveniences of old age—which impeded the works of the Evangelical life too much among the ancient men of the Gospel—and to restore the loss of youth and to recover the pristine powers and to have them again. . . . This is what all who have worked to seek the thing created for the use of humans desire, to be able to protect the corruptible body from putrefaction and to conserve it without diminution so that it may be conserved, if it is possible, in perpetuity. This is the thing that all desire naturally: never to be corrupted, nor to die.[43]

This substance that protects the body from the corruption of illness and aging is a medicine that Rupescissa names the "quintessence." He notes that many physicians have attempted to make life-prolonging medicines, but they have failed because anything derived from terrestrial materials is too corrupt to restore youth and health to the human body. Instead, he continues, one should seek something which is in itself incorruptible (if it exists in eternity), and which always makes anything that comes into unity with it uncorrupted, most of all flesh, something which nurtures the virtue and spirit of life and augments and restores it; something which digests all rawness, and reduces all that has been digested to equality, and which removes from anything all excess of any quality and restores to it any lost quality. [This root of life] makes natural humidity abound, and it manages to inflame the weak natural flame.[44]

But, he asks, how can there be a medicine accessible to humans on earth that is incorruptible and can confer its incorruptibility onto the human body?

To this we faithfully respond that it is proper to investigate this thing that is situated with respect to the four qualities of which our body is made up just as heaven is situated with respect to the four elements. Philosophers have called heaven the fifth element with respect to the four elements because heaven is in itself incorruptible and immutable and does not receive alien impressions unless it is done by the will of God. Thus, this thing which we seek is [the same] with respect to the four qualities of our body: quintessence is made incorruptible in itself, not hot and dry like fire, nor wet and cold like water, nor hot and wet like air, nor cold and dry like earth. Instead, it is a fifth element, strong against all opposing things, and incorruptible like heaven, which when it is necessary, sometimes pours in a wet quality, sometimes hot, sometimes cold, and sometimes dry.[45]

This substance, the "quintessence" or "fifth element," is naturally incorruptible and is moreover able to spread its incorruptibility to anything that comes into contact with it, including human bodies. It is able to do this because it is not made up of any of the four elements. It is instead a fifth element—inherently unchangeable and able to spread its nature to changeable terrestrial bodies.

According to the cosmology of Aristotle (384–322 b.c.e.) in *On the Heavens* and *Meteorology*, the heavens are composed of a substance that he called the "fifth body" or "ether," and which later became known as the fifth element.[46] Aristotle described the fifth element as free from change or external influence; by generating circular motion, it sustained the celestial sphere's eternality. The fifth element thus differed radically from materials composed of the four elements, which were naturally subject to "generation" and "corruption"—the coming into being and passing away that constituted change. In the writings of Aristotle's successors, the fifth element became joined to two other related ideas that proved to be significant to Rupescissa and that I discuss below: the Stoic pneuma (a vital spirit that permeated all things) and the astrological concept of a mediary that facilitated celestial influence.[47] For the sake of clarity, I refer to the Aristotelian substance of the heavens as the "fifth element" and Rupescissa's medicine as the "quintessence," although I do not mean to draw too sharp a distinction between the two.

According to Rupescissa, quintessence preserved the body and conferred incorruptibility by adding any needed "quality" to balance the body. The idea of qualities draws upon both Aristotelian natural philosophy and the medieval medical theory of "complexion" and its relation to physical and mental health.[48] *Complexio* refers to the balance of the qualities (hot, wet, cold, and dry), which resulted from the mixture of the elements (fire, water, air, and earth) in the body. The architect of complexionary medicine was the late antique physician Galen (c.130–200 c.e.), a great synthesizer of Greek medical and physical theory who became the foremost medical authority of the Middle Ages.[49]

The Galenic theory of complexion shaped both medieval conceptions of health and prescriptions of therapy. According to complexionary theory, most illness occurred because of an imbalance in the elemental qualities of the body: for instance, if a sick person suffered from a preponderance of coldness, he or she needed an infusion of qualitative heat in order to restore complexionary balance. Medieval therapeutics held that all corporeal things, including species of plants and animals, had their own particular complexions and could be used as medicines to correct imbalances. Plant and animal substances could furthermore be categorized by "grades" or "degrees" that identified the relative potency of a quality within that substance.[50] If a patient's complexion was mildly cold, he or she might be given a medicine made from a substance that was hot in the first degree; if the complexion was extremely cold, he or she might receive a medicine hot in the fourth degree, the highest concentration of the quality. A physician determined which quality was out of balance and to what degree and then prescribed a substance of the opposite quality in the corresponding degree.[51] In Rupescissa's *De quinta essentia*, medicinal substances, including flowers, roots, leaves, minerals, meats, eggs, and blood, are labeled according to their qualities and degrees in order to guide the treatment of a particular complaint:

Therefore, when you wish to know whether one of the things mentioned above is hot or wet or cold or dry: seek it in the tables of hot things and if you find it there, extract it or note it; then seek the same thing in the tables of wet things, and if you find it there, extract it. And that is how you find the complexion and grade of something. For example, should you seek quicksilver in the tables of hot things, and you find its heat in the fourth grade, and you seek the same thing in the tables of wet things, and you find it in the fourth grade, it is the highest degree of heat and humidity.[52]

According to traditional Galenic medicine, the qualities of the body were in constant flux. A sustained equilibrium—an "equal" complexion, as it was called—was desirable, but physicians thought it to be virtually unattainable. Therefore, the physician's duty was to ease continually the qualities of the body toward an elusive perfect balance. To complicate matters, each person was thought to have an innate, individual complexion that originated at the time of conception. Complexion was therefore relative. There was no single proportion that assured health for everyone; each person had his or her own particular healthy balance. This diversity was present not only among individuals but also among groups. Women were generally colder and wetter than men, and the elderly were generally colder and dryer than the young. The physician therefore had the difficult task of determining what sort of therapy was appropriate for an individual according to his or her sex, age, and unique constitution.

Closely related to the theory of complexion was that of the humors, the four bodily fluids—blood, phlegm, yellow bile (choler), and black bile (melancholy)—that were thought to make up an organism. Like all parts of the body, the humors were complexionate but dominated by a particular quality. As a result, the humors were the vehicles of the four qualities and the primary determinants of the body's overall complexion. Although the balance of qualities, rather than humors, was the chief concern of complexionary medicine, medical writers (including Rupescissa) often spoke in terms of the surplus of a particular humor, which needed to be purged in order to restore balance.[53] The dominance of a humor could make a patient vulnerable to particular afflictions; for instance, Rupescissa noted that excessive melancholy put one at risk for illness caused by demonic forces.[54]

Galenic medicine provided a theoretical foundation for Rupescissa's claim that quintessence could prolong human life. Galen suggested in his *De sanitate tuenda* that aging took place because of the loss of "radical humidity" (that is, vital heat and moisture) over the course of a lifetime; this resulted in a natural drying out of the body.[55] The "drying-out" theory was dependent on Aristotle's assertion in *De longitudine et brevitate vitae* (which was itself based on an older Hippocratic assertion) that the body became increasingly cold and dry over time. Because excessive coldness and dryness led to aging, a body that was warm and moist in perpetuity could ostensibly enjoy youth and health in perpetuity. The Aristotelian-Galenic theory of aging therefore provided a logical basis for Rupescissa's alchemy, since the quintessence balanced the complexion and compensated for dwindling heat and moisture.

But the Aristotelian and Galenic biological systems did not encourage such approaches to the prolongation of life. Aristotle drew a sharp contrast between the realms of heaven and earth: according to his theory, earth is made from the four elements and thus subject to constant generation and corruption; the heavens are made from the fifth element and thus immortal and immutable. Like all terrestrial things, Aristotle argued, human beings must inevitably pass away through drying and death. Galen accepted Aristotle's separation of celestial immortality and terrestrial mortality. He wrote that the drying out of the body was inescapable: it could be slowed down, but it could not be stopped. Such ideas about the inevitability of drying and aging also appear in Avicenna's influential *Canon of Medicine*, one of the core textbooks of scholastic medicine in the Middle Ages.[56] Although the opinions of Aristotle and Galen appear to exclude the possibility of greatly prolonged human life, Rupescissa nevertheless used them in support of his quintessence. His approach was to interpret these ideas in such a way that his conclusions flew in the face of standard naturalist theory. *De quinta essentia* relies upon Galenic medicine, but it eschews the traditional view of the body as composed of constantly fluctuating qualities that strove for—but never attained—perpetual balance. The quintessence promoted stasis in the body: it slowed or stopped the natural processes of aging, including the bodily cycles of generation and corruption and the movements of the four humors. Similarly, according to Rupescissa's analysis, the Aristotelian fifth element (and its drastic separation from the world of the terrestrial) no longer undermined the search for prolongevity but actually furnished it with its theoretical basis.

The quintessence, as Rupescissa understood it, worked to correct qualitative imbalances because "when it is necessary, [the quintessence] sometimes pours in a wet quality, sometimes hot, sometimes cold, and sometimes dry," and because "it makes natural humidity abound, and it manages to inflame the weak natural flame." The quintessence maintained the body's equilibrium of qualities and enhanced its natural humidity and warmth. This counteracted disease and the natural aging process, which drained the body's supply of moisture and heat. The quintessence could preserve the body precisely because it was not composed of any of the terrestrial four elements. Instead, it was a "fifth element" that was "incorruptible like the heavens" and thus able to do for the body what the fifth element did for the heavens. According to Rupescissa, the quintessence allowed a piece of the immortal celestial sphere to enter the mortal terrestrial sphere and confer its perfection onto human beings. This medicine bypassed Aristotelian-Galenic assumptions of the necessity of generation and corruption on earth. Instead, it imputed to humans a bit of the perfection and immortality of the superlunary world. The imagining of such a breach of the impenetrable barrier between the worlds of the terrestrial and the celestial is usually discussed in terms of Renaissance thinkers such as Marsilio Ficino and Jean Fernel, but we may see that Rupescissa expressed an early version of this doctrine in the fourteenth century.[57]

According to Rupescissa's recipe, quintessence was actually alcohol created from multiple distillations of wine that removed all traces of the four earthly elements. Because of the multiple distillations, Rupescissa contended that the quintessence was more powerful than ordinary *aqua ardens* (burning water) or *aqua vitae* (water of life), alcohol-based medicines of which medieval physicians had been aware since the thirteenth century.[58] Rupescissa also noted that quintessence could be distilled not only from wine but also from all organic and inorganic substances (including plants, minerals, and animal products) in such a way that their natural healing effects were concentrated:

God of heaven conferred such virtue to the quintessence that it extracts from all fruit, branch, root, flower, herb, meat, seed, and any species of things and from any medicinal thing, all virtues and properties and natures and effects, which the God of glory, author of nature, created in these.... and it will be a hundred times better because of the quintessence than it would be without it.[59]

Rupescissa explained that distillation extracted the active principle of a substance, thus concentrating its medicinal properties. This process resulted in a wide variety of quintessence-based drugs suited to different illnesses. Rupescissa appears to have been the first thinker to identify alcohol with an Aristotelian fifth element present in all things, an equation that proved influential, judging from the body of distillation and quintessence literature that emerged over the next two centuries.[60]

Rupescissa's claims for the quintessence point to the deep interrelationship between alchemy and medicine both in his works and in the larger therapeutic community. As early as the twelfth century, Latin intellectuals considered alchemy and medicine to be generally connected. The two disciplines resembled each other: the alchemical balance of sulfur and mercury in metals, which was thought to determine the type of metal, was analogous to the complexionary balance of qualities in the human body, which determined health. Transmutatory and medical elixirs furthermore catalyzed balance and perfection in a parallel manner, drawing another link between the two fields.

Alchemy and medicine also shared recipes and procedures. Alchemical processes, such as distillation and calcination, filtered into medicine, as did alchemical materials, such as minerals and metals.[61] The travel of ideas in the reverse direction was also common: organic materials—even human body parts and fluids, such as hair, blood, and urine—had long been used as bases for alchemical elixirs.[62] This cross-pollination between medicine and alchemy is also evident in the frequent biological analogies that appear in alchemical texts, such as the invocation of embryology to explain metal formation (more on this in chapter 6). Medical motifs occur frequently in the alchemical writings of Albert the Great, Constantine of Pisa, and Petrus Bonus, among others. Arnald of Vilanova, in particular, has been viewed by scholars as a central figure in the integration of medicine and alchemy because of his fame as a physician and the large number of alchemical texts attributed to him.[63]

In addition to medicine, Rupescissa also invoked theories of astronomy and astrology to buttress his alchemy. Astrology was an important component of medicine during this time period: because the motions of the superlunary sphere were thought to affect all terrestrial things, including the human body, physicians were obliged to learn the rules of celestial influence in order to guide their treatment strategies.[64] Astrology supposedly determined advantageous times for bleedings and other procedures, and it predicted what were called critical days, the days upon which a patient's status was likely to improve or deteriorate dramatically.[65]

Although astrology was sufficiently legitimate to be included within the curricula of the major universities by the fourteenth century, it continued to be a controversial topic. Some prominent thinkers doubted its efficacy in the hands of physicians. Arnald of Vilanova's treatise on medical astrology, *De iudiciis astronomie*, deemed most astrology to be too complicated for ordinary physicians with little training in the rigors of astronomical calculation.[66] His dismissal does not, however, point to any skepticism on his part about astrological influence: he still advises physicians to be aware of the moon's zodiacal phase when performing treatments.[67] Interest in astrology appears to have heightened dramatically in Arnald's place of residence, the Crown of Aragon, within just a few decades of his *De iudiciis astronomie*. Michael R. McVaugh speculates that the plague may be the reason for the change. The devastation of the epidemic impelled physicians to experiment with new technologies that they hoped might better counter the disease. Indeed, in 1348 the medical faculty of the University of Paris attributed to the planetary conjunction of Mars, Jupiter, and Saturn in 1345 a *pestilentia* that mutated the atmosphere, resulting in the epidemic (as well as a host of other negative phenomena).[68] In the desperate climate of the plague, astrology inspired hope and encouraged research.

It is important to note, however, that Rupescissa—like many alchemists—was not especially interested in astrological practice. Historians of science such as Anthony Grafton and William R. Newman have recently debunked assumptions that alchemy and astrology made up a coherent body of natural theory and practice in the medieval and early modern periods.[69] They show that astrological *Decknamen* (what scholars have called the symbolic code names for materials and processes used in alchemical texts) were but one type of code name among many with no particular importance; the *Decknamen* do not point to an intrinsic connection between alchemy and astrology.[70] They also point out that alchemists had little interest in astrological tables and calculations: Constantine of Pisa and Michael Scot, for instance, made few recommendations to their readers to observe astrological times except in the most general way. This characterization is for the most part true of Rupescissa. He does not include astronomical tables in his writings, nor does he advise readers to be especially attentive to the zodiac during operations. His use of astrology is chiefly general and theoretical, rather than specific and operative. In this, he fits into the mainstream of late-medieval alchemical authors on astrology. His work is distinctive, however, because of the central role that astrological *Decknamen* play in delineating the physical and metaphysical workings of the quintessence, including its symbolic connection to celestial influence.

ASTROLOGY, "PERFECTED NATURE," AND THE POWER OF GOLD

Although I have traced the ideas behind the quintessence through scholastic medicine and natural philosophy, I still have not arrived at precisely why Rupescissa thought that the quintessence was able to balance bodies on earth as the fifth element balanced bodies in the sky. Rupescissa's quintessence theory relies implicitly upon what I shall call the "perfected-nature" rationale. Rupescissa does not lay out this theory explicitly, so it will be useful to do so here. According to the perfected-nature rationale, the quintessence is the Aristotelian fifth element, a piece of the celestial sphere that channels the perfected nature of the heavens into the imperfect nature of the terrestrial world. Although Rupescissa often refers to the quintessence as "miraculous," the process was not technically "supernatural" in that it did not involve the sudden intervention of divine power into physical processes. The quintessence relied upon predictable and regular processes, but ones that were celestial and perfect rather than terrestrial and imperfect. Because the quintessence is celestial, however, it cannot be properly called "natural": it overthrows the tides of generation and corruption that are natural to the terrestrial world and its inhabitants. It is instead a form of "perfected nature"—nature borrowed from the perfected celestial sphere—that can render bodies immortal and immutable.

Indeed, Rupescissa's descriptions of the quintessence reveal a coherent ontological system that divides nature into two tiers—the corrupted and the perfected. Moreover, Rupescissa argued that alchemy allows humans to work marvels through perfected nature. He explained in *De quinta essentia* that the quintessence of blood could be used to heal wounds and cure illness because

nature perfects and has perfected this quintessence so that without any preparation it transforms blood itself from the veins immediately into flesh. And this extraordinary quintessence is the greatest thing of nature to be had, since in it is the marvelous virtue of our starry heaven and it performs the most divine miracles to the cure of nature, just as I will teach below.[71]

Rupescissa describes the quintessence as containing the "marvelous virtue of our starry heaven," that is, the properties of the celestial sphere, or perfected nature. These celestial properties perfect the quintessence so that it is able to bring about marvelous cures; this idea asserts the power of a higher order of nature that operates in the realm of alchemical medicine. We may see here Rupescissa's two levels of nature—the corrupted terrestrial nature and perfected celestial nature. He suggests that alchemy can channel the properties of the latter into the realm of the former and, as a result, the quintessence can work miraculous cures, in this case, the creation of new and healthy flesh from the quintessence of blood.

Rupescissa uses similar language in another description of the quintessence. He notes, for instance, that the quintessence possesses an odor

which must be beyond marvelous so that no worldly fragrance is able to equal it. It seems to have descended from the height of the most glorious God to such a degree that if the vase were placed in the corner of a house, because of the fragrance of the quintessence (which is a marvelous and most miraculous thing) it would attract all entering [the house] to it by an invisible chain.[72]

Similarly, the derivation of the quintessence from antimony is a "stupendous miracle," and its product "carries away the pain of all wounds and heals marvelously. Its virtue is incorruptible and miraculous and useful above all."[73] Rupescissa also claims that he will teach "in the second book [of *De quinta essentia*] to procure healthful remedies as quickly as if by a miracle."[74] Rupescissa's language appears to describe the quintessence alternately as a marvel or a miracle; it heals "marvelously," "miraculously," and "as if miraculously." Chiara Crisciani and Michela Pereira have noticed Rupescissa's use of "*quasi miraculose*" to describe the healing powers of the quintessence. They interpret his insertion of the word "*quasi*" as evidence that he "is aware that distillation does not make supernatural wonders."[75] But there is something more at work here. While Rupescissa occasionally claims that quintessence heals *as if* miraculously, there are a number of instances in which he claims that it simply *is* miraculous. His use of both "*mirabilia*" (marvelous) and "*miracula*" (miraculous) could indicate confusion about the difference between marvels and miracles—such confusion was common in naturalist texts. But this language is also indicative of Rupescissa's conception of the two tiers of nature—imperfect terrestrial nature and perfect celestial nature, the latter of which is, as he says, "beyond marvelous." This ontological distinction points to Rupescissa's belief that alchemy was a means for human beings to channel perfected nature into this world and consequently to work cures beyond the normal reach of corrupted nature.

Although Rupescissa emphasizes the heavenly nature of the quintessence, he does not suggest that one look for it in the superlunary bodies of the heavens but in the sublunary products of plants, metals, and animals. He says that quintessence is in everything: it is all around us on earth, even in our own bodies.[76] Rupescissa likely drew this idea from the long-standing theory of the pneuma, an animating principle in all things that was a part of the Stoic philosophy that came into vogue in the high Middle Ages. Historians have suggested that the concept of pneuma may lie behind twelfth-century Chartrian assertions of a "world soul," an animating force that both resided in and united all matter.[77] Furthermore, textual evidence suggests a direct link between pneuma and the Aristotelian fifth element. Frank Sherwood Taylor has identified early astrological texts that conflate pneuma with a fifth element in all things; he moreover shows that the substance was understood as a conduit for astrological influence.[78] It is possible that Rupescissa was aware of preexisting thought that identified a "quintessence" within all things on earth. An anonymous thirteenth-century text, *De generatione stellarum* (On the Generation of the Stars, formerly ascribed to Robert Grosseteste), attributes to "doctors of alchemy" a belief that "inside every natural mixed body there is a *quinta essentia*, like something that encompasses all four elements."[79] Although Rupescissa did not believe that the fifth element was a composite of the four elements, this conjunction of ideas may have prompted him to equate the vital force of terrestrial products with a quintessence and, consequently, to imagine that heavenly benefits could be harnessed by those on earth. This picture enriches the "perfected nature" rationale by suggesting how Rupescissa may have arrived at his vision of celestial powers within ordinary earthly products. Although Rupescissa was influenced by a long tradition of thought about the fifth element and pneuma, his belief that the boundary between the terrestrial and the celestial realms could be traversed was truly radical and extremely influential.

Rupescissa's recipe for the quintessence of gold, which he describes as the "sun fixed in the heavens," exemplifies this connection between the celestial fifth element and the terrestrial quintessence. It also draws upon earlier traditions that viewed the quintessence as a vehicle for astrological influence. Rupescissa writes:

> Just as the heaven above not only influences conservation in the world and creates marvelous influences not only through itself, but also through the virtue of the sun and the other stars, this heaven, the quintessence, wishes to be adorned with the marvelous sun, which equals it in splendor and incorruptibility. In such a sun, fire itself is not able to act in such a way that it corrupts itself. And I say to you in true charity and good conscience, that this sun—illuminated, splendid, and uncorrupted by fire, which pours out incorruptibility and the root of life in our body in such a way as is possible, as I have explained above, and which was created in order to adorn our heaven and to augment the influence of the quintessence—can be captured in the hand. The King of glory put it in the power of mortals and because of the charity of God, which speaks to Evangelical men, I reveal it to you with its proper and intelligible name: this is the gold of God, which is made from the Philosophers' stone. ... The sun certainly is the son of the sun of heaven, from which the Philosophers' stone is composed. For it is generated from the influx of the sun in the viscera of the earth, and the sun with its influence, grants to it its nature and color and incorruptible substance.[80]

In this passage, Rupescissa identifies the metal gold with the "sun" and the quintessence with "heaven" by using *Decknamen*, symbolic code names for alchemical substances or processes. The connection between the metal and the sun allows Rupescissa to offer a theory of astrological influence to explain the effects of quintessence and gold upon the human body. According to the standard Aristotelian cosmological view, the movements of the celestial bodies cause generation and corruption on earth by prompting changes in the terrestrial elements and their primary qualities.[81] Rupescissa compares the influence of the quintessence and gold on the human body to the effects of the celestial fifth element and sun on the terrestrial world. According to this analysis, the quintessence is a mediary that provides stability to bodies and facilitates "astrological" influence from the "sun." Recourse to astrology thus provides a theoretical basis for the efficacy of quintessence, as well as for its combination with gold.

The parallels that Rupescissa draws between gold and the sun are quite traditional. Gold played a part in antique and medieval conceptions of the human as a microcosm, a miniature reproduction of the world that contained all of its elements. According to this model, just as the sun delivered its vital force to the universe, the mineral sun—gold—delivered vital heat to the human body.[82] Gold had not only a symbolic connection to the sun but also a more direct one, according to Rupescissa. He theorized that the heat of the sun led to the generation of gold in the earth's interior. During the process of generation, he writes, the sun transfers its properties, including its color and incorruptibility, to the metal. As a result, gold is able to withstand great heat without being damaged. This process parallels Rupescissa's understanding of gold's effect on the quintessence, and on those who consumed the quintessence, who also absorbed some of the properties of gold. Medieval mineralogical theory held that gold was an "equal" body (a metal comprising perfectly balanced constituent elements). Alchemists and physicians thus frequently suggested that a medicine made from gold might create in the human body a perfectly balanced complexion.[83] The genealogy from the celestial sun to the human body is built upon a sort of contagious perfection. The sun transfers its properties to the metal through generation, and the resulting metal conveys those properties to the human body by manipulating qualitative balance. Medicinal gold also indicates an important convergence of Rupescissa's transmutatory and medical alchemy. *De quinta essentia* praises the "gold of God" as a superlative base for quintessence and, as the passage above indicates, this gold is synthesized by means of a philosophers' stone.[84] Alchemical gold serves as a bridge between Rupescissa's *Liber lucis* and *De quinta essentia*, since the recipes in the former are required to achieve the cures of the latter.

Rupescissa's interest in potable gold as a medicinal substance also points to the context of his writings in the crisis of the Black Death, during which potable gold became a topic of great medical interest. In the midst of the plague, a number of healing professionals became involved in alchemy and its search for a universal medicine, a single substance that could cure any illness, including plague. Prior to the Black Death, medieval confidence in the efficacy of medicine was high, but expectations were frustrated by the advance of the disease.[85] The plague did not conform to the widely accepted principles of humoral medicine, and it afflicted people of every humoral type. Katharine Park has argued that the high mortality associated with the plague led to changes in the practice of conventional medicine when it failed to stop the epidemic. Chiara Crisciani has suggested that medicine, which was generally ineffective against the plague, looked eagerly to alchemy for innovative methods.[86]

Exotic remedies involving herbs, minerals, and even pearls and gems can be found in therapeutic writings of the mid-fourteenth century, and the wealthy had access to a wide variety of expensive ingredients.[87] Recipes for counter-plague elixirs varied, but they were often produced by alchemical methods, and potable gold was thought to be especially effective. Gentile da Foligno, Thomas of Bologna, and "Solemnis Medicus" all offered recipes for potions involving gold; some of these potions were produced alchemically, although by different processes than that suggested by Rupescissa. Thomas of Bologna, in particular, voiced interest in using the perfectly balanced complexion of gold to impose perfect balance upon the human body and thus cure the plague (among other diseases). Other writers, including Maino de' Maineri, proposed wine-based alchemical remedies, although such thinkers did not claim these remedies to be either "quintessences" or pieces of the celestial sphere. A number of practitioners and theorists also searched for universal antidotes, including "theriac," a long-discussed remedy rumored to counter any poison.[88] Over time, Chiara Crisciani and Michela Pereira argue, practitioners shifted from prescribing a variety of remedies for the plague to emphasizing a single and universal remedy that would cure any illness. Rupescissa's writings on quintessence and gold thus fit into this contemporary discussion of potable gold, universal remedies, and alchemy in the writings of physicians and theorists. Rupescissa attacks the ideas of a certain "Bernar[d]us magnus" on the causes of gold's medicinal efficacy, demonstrating that he was well familiar with others' conclusions about alchemy and potable gold. This suggests that Rupescissa was reading the works of other writers and was fully involved in the conversation about elixirs that emerged from the climate of the plague.[89]

Rupescissa's recipes for the quintessence contain a sophisticated mix of information culled from various bodies of knowledge, including natural philosophy, medicine, and astrology. Rupescissa situated his project firmly within the scholastic Aristotelian-Galenic inheritance of natural philosophy and medicine, but he extended that inheritance in dramatically new directions and arrived at a radically new conclusion. Rupescissa searched for remedies both inside and outside the standard works of medicine and natural philosophy taught in European universities. He thereby encountered in his studies the scientific theories of several significant thinkers who predated him—Roger Bacon, Arnald of Vilanova, and Ramón Llull. Their ideas guided Rupescissa in his application of naturalist thinking to eschatological problems and set the stage for his alchemical innovations. The next chapter will survey these sources for Rupescissa's theories in order to demonstrate how they provided for him both a conceptual foundation and a point of departure.

Five

ARTISTS AND THE ART

In the fourteenth century, an aspiring alchemist had at his or her disposal a considerable array of texts to comb for the secrets of the art. Some texts provided only instructions for the synthesis of gold and silver; some offered remedies for the preservation of human health and youth; some even hinted at the great social change that would result from limitless wealth and youth. It is perhaps no wonder that the Franciscan and Dominican orders promulgated a series of bans on alchemical practice in the late thirteenth century, or that John XXII, who was skeptical of transmutation but worried about counterfeiting, forbade the minting of coins from alchemical gold in the early fourteenth.[1] Beyond the practical concerns of these lawmakers lay larger problems: if alchemical gold was freely manufactured, what might happen to material wealth and the power that derived from it? What might a society of extremely long-lived and healthy humans be like? Questions such as these fueled the production of alchemical texts throughout the Middle Ages and beyond. Such texts held considerable sway over the imaginations of alchemical readers, including John of Rupescissa.

Although Rupescissa's particular formulation of alchemy and apocalypticism was highly original, his work was nevertheless informed by the earlier writings of a number of naturalists, some of whom also had a profound interest in apocalypticism. Works by Roger Bacon (1214–1292), Arnald of Vilanova (c. 1240–1311), and Ramón Llull (c.1232–1316) proved foundational for Rupescissa's apocalyptic alchemy. Bacon promoted the study of nature in his *Opus* trilogy, an ode to the power of knowledge to save the church. Arnald composed texts on both eschatology and science, along the way advancing medical therapeutics, apocalyptic hermeneutics, and the Franciscan Spiritual cause. Ramón Llull compiled knowledge of the natural world as a means to convert the infidel and recover the Holy Land—both goals with eschatological dimensions. While there are clearly differences between the medical treatises of Arnald of Vilanova and the *ars combinatoria* manuals of Ramón Llull, they are alike in an important respect: each promotes naturalist inquiry for spiritual purposes. In addition, Arnald and Llull's names became attached to a body of spurious alchemical writings that exercised a profound influence on Rupescissa, who admired Arnald and Llull and viewed himself as following in their footsteps. In order to understand Rupescissa's apocalyptic alchemy, it is crucial to place it within the context of the sources that shaped it.

Roger Bacon, Arnald of Vilanova, and Ramón Llull all provoked skepticism and even outright hostility from many of their contemporaries. Bacon and Arnald—both outspoken men with abrasive personalities— made enemies powerful enough to engineer Arnald's trials for heresy and Bacon's eventual imprisonment. Llull's ingenious system of the *ars combinatoria* was, to some extent, overshadowed by his reputation as an odd, countercultural figure, leading to his frequent dismissal as "Ramón the fool." Rupescissa's accomplishments were dependent upon the writings of these men, whose interests and travails paralleled his own, yet Rupescissa also diverged sharply from his predecessors in his theoretical and practical conclusions. Only in Rupescissa's writings do we find a coherent, systematic coupling of quintessence alchemy and Joachite apocalypticism. Among these authors, only Rupescissa points to alchemy as a solution to a crisis that is wholly apocalyptic.

ROGER BACON AND THE SEARCH FOR THE EQUAL BODY

Attempts to prolong life through alchemy had a long history by the time of John of Rupescissa's career in the mid-fourteenth century, and the papal court—where Rupescissa was imprisoned as he wrote his alchemical treatises—was a center for the exploration of such ideas. The historian Agostino Paravicini Bagliani has painted a vivid portrait of a late medieval papal curia headed by elderly popes and cardinals searching desperately for the fountain of youth.[2] Exploring the preservative qualities of alchemy became something of a trend in this climate. The fabrication of alcohol, or *aqua ardens*, was well known to the curia at least by the fourteenth century. Cardinal Vitalis of Furno (d. 1327 in Avignon) was the author of a treatise on healing waters, *Pro conservanda sanitate*, and, in 1330, Pope John XXII reportedly gave money to his doctor, Bishop Gaufré des Isnards, to buy an alembic for distillation. According to the testimony of Arnald of Vilanova, who was also Boniface VIII's physician, the cardinals of the curia consumed potable gold for their health with regularity.[3]

Several generations earlier, in the mid-thirteenth century, Pope Clement IV (1265–1268) entertained the theories of one of alchemy's most vocal proponents, Roger Bacon, who wrote for him a series of works on the wonders of naturalism.[4] Bacon's program of study included mathematics, optics, medicine, and astronomy, among other disciplines, but he deemed alchemy to be especially crucial. Alchemy was the cornerstone of naturalist knowledge, Bacon argued, since it dealt with the generation of things from the four elements. Perhaps most significant, he believed that medicines manufactured by alchemical methods could allow human beings to extend their lifespans to extraordinary lengths.[5] He treated the subject in five different works, *Epistola de secretis operibus artis et naturae* (Letter on the Secret Works of Art and Nature, 1249–1252), *Opus maius* (Great Work, 1266–1268), *Opus minus* (Small Work, 1267), *Opus tertium* (Third Work, 1267–1268), and *In libro sex scientarum in 3° gradu sapientie* (In the Book of the Six Sciences in the Third Grade of Wisdom), a fragmentary compendium of the ideas contained within the *Opus* trilogy.[6] Bacon's imagination was captured by his encounter with a Pseudo-Aristotelian text known as the *Secret of Secrets*, and he authored an introductory treatise and annotated edition of the text.[7] The alchemical content of the *Secret of Secrets* exercised such a powerful influence on Bacon that he thereafter considered alchemy to be a technology with great potential to improve human life.

According to Bacon, contemporary human beings led much shorter lives than was natural. He mused that the biblical patriarchs lived for one thousand years or even longer, and he indicated that such was the natural lifespan of all people.[8] Bacon blamed the current brevity of human life on two factors: first, a decline in morality and, second, the long-standing neglect of the regimen of health. Human beings passed these poor habits onto their progeny, perpetuating the problem:

[F]or these two reasons the longevity of men has been shortened contrary to nature. Moreover, it has been proved that this excessive shortening of the span of life has been retarded in many cases, and longevity prolonged for many years by secret experiments. Many authors write on this topic. Wherefore this excessive shortening of life must be accidental with a possible remedy. . . . The proper regimen of health, therefore, as far as a man can possess it, would prolong life beyond its common accidental limit, which man because of his folly does not protect for his own interest; and thus some have lived for many years beyond the common limit of life. But a special regimen by means of remedies retarding the common limit mentioned, which the art of directing health does not exceed, can prolong life much further.[9]

This passage makes two important arguments: First, extraordinarily long life is actually *natural* for humans and short life *unnatural*; the current situation is thus correctible. Second, humans can regain natural longevity by following a regimen of health, which includes special remedies that retard aging. Bacon's approach to health expands upon Galenic medicine by not only regulating hygiene to achieve the current normal lifespan but using therapeutics to extend it much further.

Bacon notes that God gave to humans and animals knowledge of special medicines that can cure disease and protect the body from "cold" and "drying up."[10] These medicines include a variety of odd substances such as the flesh of serpents, stags, and other supposedly long-lived creatures.[11] Bacon was particularly interested in substances that induced an "equal complexion," that is, a rightly proportioned (though not necessarily equal) mixture of elements in a body. Among the substances that conferred equality was, as one might expect, the alchemical elixir. Bacon explains: "For that medicine that would remove all impurities and corruptions of a baser metal, so that it should become silver and purest gold, is thought by the wise to be able to remove the corruptions of the human body to such an extent that it would prolong life for many ages."[12] Bacon offers various methods for creating an equal body through alchemical techniques. For instance, he writes in his commentary on the *Secret of Secrets* that one should separate human blood into its constituent elements and then recombine these materials in "equal" proportions to produce an elixir.[13] According to Bacon, this substance could transmute base metals into the incorruptible state of gold or human bodies into the incorruptible state of youth and health.

No one before Bacon had argued that the corruption of the body was due to "accidental," and thus remediable, factors.[14] This was a revolutionary departure from Aristotelian and Galenic theory, as well as from the Christian theological view of mortality, all of which held that the human body was naturally corruptible and ephemeral, at least until its resurrection.

Bacon moreover differed from his contemporaries by applying alchemical "medicines" to human bodies, as well as to metals. He was among the first Europeans to discuss the healing powers of potable gold and to use the Arabic-derived word "elixir." Historians have not yet determined the path of the elixir idea to the West, but it seems likely that Arab naturalist authors were the source. Gerald J. Gruman points to the Jābirian elixir, the "medicine" that healed metals, as the probable basis for Bacon's alchemical theories; he further traces the Jābirian model to earlier Chinese Taoist ideas about the prolongation of life.[15]

John of Rupescissa likely owes a debt to Baconian alchemy, which he may have encountered either directly or through a text such as the *Rosarius philosophorum* (*Rosary of the Philosophers*).[16] Like Bacon, Rupescissa claimed that an "equal" (i.e., rightly proportioned) complexion was the key to health and youth and that potable gold and alchemical elixirs could play a central role in achieving it. Rupescissa, however, formulated a very different explanation of disease and aging, one that pointed him away from Baconian therapeutics and toward the quintessence as a means of preservation. First, Bacon wrote that immortality was the natural condition of humans, and one that they would still enjoy if not for sin.[17] Moreover, he found in the antediluvian myth evidence that even postlapsarian human beings could live for one hundred fifty years or longer by regulating the nonnaturals and using alchemical medicines. Bacon's goal was to use alchemy to return the body to an earlier state of natural balance and quasi immortality. He noted that certain people appeared to have achieved this state even in his own day; for instance, he pointed to a contemporary English holy woman who reputedly survived without eating or excreting. Bacon speculated that this woman had obtained the natural qualitative balance that led to bodily incorruptibility.[18]

While Bacon argued that incorruptibility was the natural condition of the human body, Rupescissa adopted the more orthodox view that humans were by nature corrupt. He claimed that the natural physical discomforts of disease and old age could be remedied through alchemical medicines that worked, as he said, "against the infelicities of nature."[19] Rupescissa therefore did not hope to return to some previous state of nature but to *overthrow* corrupted nature. This forced him to emphasize the need for something outside of ordinary nature to bring about incorruptibility and longevity in the human body, something that could oppose nature's "infelicities." Rupescissa's quintessence alchemy constitutes a significant departure from Bacon's solution.[20] It is possible that Rupescissa's theory of the quintessence was influenced by Bacon's statement in the *Liber sex scientarum* that "burning mirrors" could concentrate and project rays of beneficial celestial starlight onto equal bodies created by alchemy. These bodies could then transfer the power of the stars to humans by multiplying their species even when the stars were no longer in a beneficial position.[21] Although this process works by harnessing the power of the celestial bodies, it is nevertheless a vastly different system, relying on astrological influence and species propagation, not on the entrance of "perfected nature" into terrestrial nature.

Rupescissa also refrains from Baconian claims of extraordinary longevity for human beings. He instead emphasizes the maintenance of health and youth up until the end of life without specifying a particular lifespan. He displays little interest in extending alchemical cures to all people, instead focusing on a limited number of evangelical men slated to play key roles in an apocalyptic drama. As we have seen, Rupescissa expected many people to die in the apocalyptic plagues, famines, and wars that would accompany the advent of Antichrist. Nor did he think that such tremendous loss of life was completely negative: the purge of society brought about by Antichrist would actually improve Christendom. The distinctions between Bacon's and Rupescissa's thoughts were largely a product of their different aims. While apocalyptic concerns were an important element of Bacon's alchemy, as I shall discuss below, they are not the predominant theme; instead, alchemy was part of an overarching program of naturalist research. Rupescissa, in contrast, placed all naturalist inquiry within an apocalyptic framework. Rupescissa's disagreements with Bacon make sense within such a context: there would be little need for extreme longevity for all humans when, as he claimed, the earthly paradise of the third state was about to supersede the current corrupt age.

For Roger Bacon, images of a rejuvenated and quasi-immortal human body were intrinsically connected to the resurrection of the flesh in Christian doctrine. He argued that the equal complexion achieved through alchemical therapeutics was the same as the principle of immortality in the postresurrection body:

For this condition will exist in our bodies after the resurrection. For an equality of elements in those bodies excludes corruption forever. For this equality is the ultimate end of the natural matter in mixed bodies, because it is the noblest state, and therefore in it the appetite of matter would cease, and would desire nothing beyond. The body of Adam did not possess elements in full equality, and therefore the contrary elements in him acted and were acted on, and consequently there was waste, and he required nourishment. For this reason he was commanded not to eat the fruit of life. But since the elements in him approached equality, there was very little waste in him; and hence he was fit for immortality, which he could have secured if he had eaten always the fruit of the tree of life. For this fruit is thought to have elements approaching equality; and therefore it was able to continue incorruption in Adam, which would have happened if he had not sinned. Wise men, therefore, have striven to reduce the elements in some form of food or drink to an equality or nearly so, and have taught the means to this end.[22]

This passage makes some startling connections between alchemy and the resurrection body. First, Bacon writes that human beings can obtain the state of equality, which is also the harmony and incorruptibility of the postresurrection body, in this world, before the end of time (although only in a partial and imperfect way, as he notes elsewhere).[23] Second, Bacon seems to equate the alchemical "equal medicine" with the fruit of the tree of life, which possesses the elements in an almost perfect balance and which could transfer its incorruptibility to Adam's body.[24] Third, he suggests that humans who achieve equality in this lifetime through artificial means are especially "fit for immortality." And, indeed, Bacon elsewhere explains that at the moment of resurrection the body is not able to achieve incorruptibility except through equality.[25] Bacon makes quite a claim for the power of alchemy: through its practice, humans could gain the benefits of the postresurrection body before the resurrection and, furthermore, prepare themselves in advance for eternal life.

Did Rupescissa think that the body balanced by quintessence was the same as the resurrection body? We cannot know for certain since he does not discuss the resurrection explicitly. But it seems more than coincidence that Rupescissa devotes such attention to a substance that imparts incorruptibility and quasi immortality to bodies at the same time that he predicts the millennium, the one-thousand-year earthly reign of the saints. Moreover, Rupescissa repeatedly refers to the quintessence as "our heaven," revealing a link between, on the one hand, the final incorruptibility that theologians argued would happen only after the resurrection and, on the other hand, the medicinal benefits of the quintessence in the present earthly world. How does this fit into his millenarian predictions of peace on earth, a "heaven" of sorts for humankind?

As we have already seen, Rupescissa's *Vade mecum* includes powerful images of sinful lords and prelates undergoing vicious bodily dismemberment and corruption. Subjects tear apart their oppressors, who are then devoured by worms or the "mouths" of swords. Rupescissa emphasizes bodily division and consumption: in each case, the guilty are rent or sliced into pieces, and then (at least metaphorically) digested by their subordinates.[26] This scene is particularly significant in light of its contrast to the evangelical men—incorrupt, undivided, and untroubled by any threat of putrefaction, digestion, or bodily change until the time of their predetermined deaths. Rupescissa appears to indicate that the rewards and punishments promised in the next world will occur here on earth and within time. While the sinful will experience the bodily fragmentation of hell, the saintly evangelical men will be the picture of bodily integrity and incorruptibility, making an easy transition from this era to the earthly paradise of the third state. Rupescissa's spiritual vision is markedly corporeal: the human body extends from contemporary society into millennial society, combining this world and the next into one coherent earthly history. This narrative makes clear a key aspect of Rupescissa's thought. His apocalypticism and his alchemy are both concerned with the same subject—the defense of the Christian body in this world and its safe transition into the next. Furthermore, as I shall argue in the next chapter, it was Joachite apocalypticism, with its potent idea of the third state, that prompted Rupescissa to imagine the alchemical quintessence as a piece of heaven available here on earth and within time.

Eschatology also plays a part in Bacon's alchemy. He devotes several substantial portions of the *Opus maius* and *Opus tertium* to concerns about last things, particularly Antichrist. The problem of predicting and combating Antichrist functions as a leitmotif in Bacon's discussion of what he called *scientia experimentalis*, a body of knowledge encompassing optics, medicine, mathematics, applied astronomy, and alchemy. According to Bacon, each of these sciences could offer some form of support to the church in the last days. Bacon professes a particular affinity for astronomy/astrology because, as he argues, this science is the best for reconciling contradictory calendars and predicting Antichrist's arrival.[27] Bacon also praises the prophecies of the "Sibyl and of Merlin, Aquila, Seston, [and] Joachim," which can be scrutinized to gain "greater certainty regarding the time of

Antichrist."[28] Here, Bacon argues for the reconciliation of both textual and natural observational clues in order to predict more accurately Antichrist's plans. As he puts it, any theologian "must know the things of this world if he ought to know the sacred text."[29]

Based on his understanding of celestial phenomena and textual prophecy, Bacon concludes that an Antichrist-led army of natural philosophers will soon attack Christendom, necessitating countermeasures by a parallel group of Christian natural philosophers. It was therefore of utmost importance that Christians embrace training in *scientia experimentalis*:

For this science teaches how wonderful instruments may be made, and uses them when made, and also considers all secret things owing to the advantages they may possess for the state and for individuals. And it directs other sciences as its handmaids, and therefore the whole power of speculative science is attributed especially to this science. And now the wonderful advantage derived from these three sciences in this world on behalf of the church of God against the enemies of the faith is manifest, who should be destroyed rather by the discoveries of wisdom than by the warlike arms of combatants.[30]

According to Bacon, the church could prevail by means of technologies based upon the principles of nature. For instance, Bacon speculates elsewhere that the church could learn to alter the atmosphere of the earth, affecting human complexions and making conditions favorable for the defeat of its enemies.[31] He continues in the *Opus maius*:

Antichrist will use these means freely and effectively, in order that he may crush and confound the power of this world. . . . Moreover, the church should consider using these methods against unbelievers and rebels, in order that it may spare Christian blood, and it should do so especially because of the future perils in the times of Antichrist, which with the grace of God it would be easy to meet, if prelates and princes promoted study and investigated the secrets of nature and of art.[32]

Bacon offered these warnings to Pope Clement IV in part to prepare the church for Antichrist's assault. In recent years, a scholarly debate has grown up around the question of whether Roger Bacon was a Joachite—a follower of Joachim of Fiore and his apocalyptic theology. Stewart Easton has concluded that Bacon was an unequivocal Joachite; Marjorie Reeves and E. Randolph Daniel have urged caution in applying such a label, instead arguing that Bacon was merely influenced by Joachite thought.[33] More recently, David C. Lindberg has refocused on Bacon's apocalypticism—whether Joachite or merely Joachite influenced.[34] Unlike Rupescissa, Bacon includes no explicit references to Joachite structures such as the spiritual men or the third state in his writings. It therefore seems unlikely that scholars will come to a definitive conclusion about Bacon's Joachism or lack thereof. But he was clearly guided by a sense of impending apocalyptic crisis. We may see in Bacon's writings an early confluence of apocalyptic thought and natural scientific inquiry, a combination that would continue to find enthusiasts for decades to come.

ARNALD OF VILANOVA'S MEDICINE FOR CHRISTENDOM

A convergence of naturalist and apocalyptic thought appears not only in the mid-thirteenth-century writings of Bacon but also in the work of Rupescissa's near contemporary, Arnald of Vilanova (c. 1240–1311), the controversial physician and reformer whose writings made a profound impression upon Rupescissa's prophecy and alchemy. Arnald of Vilanova was among the most famous physicians of his day: he achieved renown both as a master of medicine at the University of Montpellier and as the personal physician of kings and popes.[35] Like Rupescissa, Arnald was a man of both naturalist and religious leanings. He provided for his royal and papal patrons not only medical but also (sometimes unsolicited) spiritual advice. During his early career at Montpellier, Arnald composed a number of influential medical texts on topics ranging from epilepsy to pharmacology to love sickness. Later in his life, he aligned himself with the Franciscan Spiritual movement and Joachite apocalypticism. During this period, he composed his prophetic tract *De tempore adventus Antichristi*, which Marjorie Reeves, Harold Lee, and Giulio Silano have described as the first written record of his Joachite affiliation and a document of central importance to his religious career.[36] Arnald subsequently became an ardent spiritual reformer and an outspoken defender of the Beguins, for whom he provided vernacular translations of Peter Olivi's Latin writings, as well as his own tracts of encouragement, the *Alia informatio beguinorum* and *Informatio beguinorum seu Lectio Narbone*.[37] He followed his *De tempore adventus Antichristi* with a series of prophetic works on the advent of Antichrist—a theological presumption that twice placed him in danger of his life. (He was famously condemned in 1301 by Boniface VIII, who thereafter suffered a fortuitous illness and submitted himself to Arnald's medical care. Arnald then reentered the pope's good graces.)[38]

It would hardly be possible to overestimate the importance of the Arnaldian alchemical corpus to Rupescissa, who cites "Master Arnald" in his *Liber lucis* frequently and reverentially. The state of current scholarship is such that it is not possible to determine which, if any, alchemical works are the genuine works of Arnald. Juan A. Paniagua dismisses all such alchemical works as spurious, maintaining that Arnald did not believe alchemical transmutation to be possible. Lynn Thorndike and Antoine Calvet, in contrast, accept a few Arnaldian works as genuine, including the *Rosarius philosophorum*.[39] Calvet argues that Arnald was not generally hostile to alchemy, and, indeed, many alchemical precepts appear in Arnald's standard medical works. But Joseph Ziegler, the author of a monograph on Arnald's medical and religious thought, accepts Paniagua's thesis and includes no alchemical works in his study.[40]

There is no doubt that Arnald of Vilanova engaged in research on the prolongation of life. He wrote a medical treatise on aging, *De humido radicali*, and he inserted the prolongevity tract *De retardatione* nearly word for word into his *De conservanda iuventute et retardanda senectute*.[41] There is also some independent evidence that the genuine Arnald practiced transmutatory alchemy—although the evidence dates from the 1340s and 1350s, when Arnald had already begun to acquire his posthumous reputation as an alchemist.[42] A witness attached to the papal court, the lawyer Giovanni d'Andrea, remarked in his *Additiones* to the *Speculum iudiciale* of William Durant that apart from being the greatest doctor and theologian in the Roman curia, Arnald was also a "great alchemist" who manufactured pieces of alchemical gold.[43] An alchemical work entitled the *Correctio fatuorum* claims that Arnald successfully healed a pope by using alchemical gold; other physicians who had attempted to treat the pope had failed. The scholar Nicolas Weill-Parot has recently articulated the many links between Arnald's medical, astrological, and talismanic practices, on the one hand, and the principles of alchemy, on the other.[44] For our purposes, however, whether or not the historical Arnald wrote about or practiced alchemy is not as important as John of Rupescissa's belief that alchemical works circulating under Arnald's name were genuine. The ideas contained within them about alchemy, alcohol, and health very likely inspired components of Rupescissa's quintessence.

Arnald's genuine textual production is so enormous that it cannot be treated here in detail. I will merely touch upon a few themes that resonate with the later ideas of Rupescissa. Under the patronage of Frederick III of Sicily (1296–1337), whom he recognized as an apocalyptic "God-elected king," Arnald spearheaded a program of spiritual and societal reform to prepare the world for the last days. In order to reshape Sicily into a perfect Christian society, Arnald recommended the conversion of the Jews, the building of special hospitals and housing for the poor, and the institution of evangelical schools to train men (and some women) for their apocalyptic mission of preaching.[45] Study of the natural world was also a part of Arnald's reform program. In the *Allocutio christiani*, a treatise written for Frederick III, he promoted a rationalist approach to nature that would ultimately lead to an appreciation of divine dignity:

God gave reason to [man] so that by considering intelligible things from sensible things, he would know to perceive the excellences or dignities of God through it. This is proven through the sensible things so that thus through the cognition of oneself in Him, as much as is possible in the present life, and through the cognition of his dignities, his soul would warm toward loving God and because of that love, he would praise Him. Man, however, knows God in the present life, first of all through his creatures. When one considers these according to their origin and multitude and magnitude and beauty and order and operation, these things shine: immeasurable power, inexplicable wisdom, interminable goodness of the Creator.[46]

Arnald goes on to argue that God has implanted providentially the meaning and order of life in the natural world so that humans may decipher it through reason. Naturalist knowledge was thus for Arnald indispensable to the pursuit of theology. Michael R. McVaugh has suggested that Arnald developed this interpretation of nature and human knowledge in order to justify his theory of progressive human understanding (a version of which we have already seen among the apocalypticists I discussed in chapter 3). Arnald claimed that all human knowledge contained the elements for prophecy. Thus, as human knowledge expanded, so too did humanity's potential for understanding prophecy. Fields of naturalist knowledge were among these paths to divine truth and the perfection of the human soul.[47]

Arnald also viewed his spiritual and medical vocations as compatible, as Joseph Ziegler has argued in a persuasive study. According to Ziegler,

Arnald extended his medical approach to spiritual pathologies: he believed that because medical knowledge was divinely inspired, physicians could treat the spiritual ills of society as well as the physical ills of the individual patient. He portrayed Christ as a physician and suggested that medical practitioners might have special access to divine knowledge.[48] Arnald's call to study the natural world was doubtless wrapped up in spiritual concerns, and I would suggest that this pairing of approaches had great consequences for John of Rupescissa. Arnald's prophetic writings, especially the *De mysterio cymbalorum ecclesiae* and *De tempore adventus Antichristi*, were among the sources for Rupescissa's exegetical interpretations. There is no reason to assume that Rupescissa did not also take note of Arnald's thoughts on the place of naturalist knowledge within an apocalyptic scenario.

As I noted above, Arnald has been credited with the authorship of a large number of alchemical works that he likely did not write. Some of these texts were central to the formation of Rupescissa's alchemy.[49] Antoine Calvet suspects that the Pseudo-Arnaldian text cited by Rupescissa in the *Liber lucis* as the "*Tractatus parabolicus*" is related to the famous *Rosarius philosophorum* attributed to Arnald; moreover, Rupescissa credits both "Arnald of Vilanova" and a "*Rosarius*" in his text, although he cites the two as different authorities.[50] Several ideas from the Arnaldian *Rosarius* also surface in Rupescissa's writings, perhaps indicating a direct link between the works. Like Rupescissa's *Liber lucis*, the *Rosarius* describes the traditional separation of the four elements from the philosophers' stone by distillation and calcination, a process that purifies it of terrestrial matter. Less traditional is the author's assertion that the philosophers' stone "conserves health, strengthens the mind and virtues, rejuvenates old men, expels all illness.... This medicine is certainly to be sought after above all other medicines or wealth of this world, since he who has it has an incomparable treasure."[51] While variations on these statements appear in other alchemical texts, the common elements of the *Rosarius* and Rupescissa's alchemy, along with Rupescissa's explicit mention of the author and title, suggest that Rupescissa read either this treatise or one closely related to it.

Another text that has been attributed to Arnald, the *Tractatus de vinis* (Treatise on Wines), was also a likely source for Rupescissa's ideas. This text, probably composed from 1309 through 1311 for Robert of Naples, is a paean to the virtues of medicinal wines.[52] A number of the healing properties with which Rupescissa credits the quintessence are found here in some form. The author cites medicinal wines as a cure for leprosy, paralysis, and quotidian and quartan fevers—all illnesses treated by the quintessence in Rupescissa's work. The author of *De vinis* also proposes that wine has the ability to prolong human life. When infused with rosemary, it "preserves one from all evil ulcers and consumes bad humor and melancholy. It strengthens by its own virtue the substance of the heart and thus keeps people young. And perhaps the body of one who uses it permanently will not decay."[53] Here we see prolongevity claims joined with the healing powers of wine, a juxtaposition that Rupescissa may have adapted for *De quinta essentia*. A recipe for potable gold, one of Rupescissa's most cherished remedies, also appears in *De vinis*. The author notes that wine prepared with gold leaf strengthens the heart, purifies the blood, and preserves youth.

The author of *De vinis* used a different method for preparing medicinal wines than Rupescissa recommended in *De quinta essentia*. While *De vinis* suggested that herbs or other medicinal materials should be steeped in hot or cold wine, Rupescissa instructed readers to distill herbal and animal products repeatedly to extract their quintessences. These methods differ substantially: although the author of *De vinis* deemed the infusion of healing properties into wine to be sufficient, Rupescissa believed one had to release a heavenly element from within the materials through distillation. The author of *De vinis* mentions distillation only briefly and moreover does not recommend minerals besides gold as medicinal additives to wine. Nevertheless, *De vinis*'s enthusiasm for wine may have directed Rupescissa toward the formulation of a related remedy.[54]

Another Pseudo-Arnaldian text focuses upon distilled alcohol, or *aqua vite*, in detail. This treatise, the *Tractatus de aqua vite simplici et composita*, appears to function as a bridge between the works of Arnald and Rupescissa by making the first gesture toward Franciscan or Joachite apocalypticism within an alchemical-medical treatise. One manuscript copy of the text bears a brief dedication to the "evangelical poor." The historian Antoine Calvet notes, however, that *De aqua vite* does not integrate evangelical poverty or Joachite apocalypticism into its overall message.[55] There is no further mention of apocalypticism or the role of *aqua ardens* in the last days in the rest of the text. It is possible that *De aqua vite* was known to Rupescissa, and this short snippet of text may have helped to forge a link in his mind between distilled remedies and the fate of his own evangelical men.[56] But Rupescissa diverges from the treatise by arguing that the quintessence, which must be distilled multiple times, is something more than mere alcohol. According to Rupescissa, the "quintessence both is *aqua ardens* and is not *aqua ardens*."[57]

Although Rupescissa and Pseudo-Arnald shared dual interests in alchemy and evangelical poverty, Rupescissa's theory of the quintessence and its relationship to apocalyptic crisis has no parallel in the Pseudo-Arnaldian writings. Well-known Pseudo-Arnaldian alchemical texts, such as the *Novum lumen* (New Light), *Flos florum* (Flower of Flowers), *Testamentum* (Testament), *Semita semitae* (Path of Paths), and *Epistola super alchimia ad regem neapolitanum* (Letter on Alchemy to the King of Naples) are chiefly concerned with transmutatory alchemy. Their structure and content thus differ significantly from Rupescissa's *De quinta essentia*, which was focused on medical therapeutics. Nevertheless, a few remarks from within the Pseudo-Arnaldian corpus are noteworthy. First, Pseudo-Arnald introduces the idea of a "quintessence" in the *Quaestiones tam essentiales quam accidentales*. But for the author quintessence is the singular substance of the philosophers' stone, rather than a group of alcohol-based remedies.[58] Second, the author of the Pseudo-Arnaldian *Testamentum* writes that the stone is "similar to the nature of the heavens because of its nobility, or substance, and the subtlety that it contains within it, and for these reasons they call it *spiritus*."[59] Although this brief statement links the philosophers' stone to heaven, the author does not extend the comparison, nor does he suggest that the stone's heavenly aspects have medicinal value. Pseudo-Arnald elsewhere discusses medicinal alcohol, but he always names it *aqua vite* (water of life) or *aqua ardens* (burning water), not *nostrum caelum* (our heaven) or *caelum humanum* (human heaven). Since the Pseudo-Arnaldian alchemical corpus does not propose any ontological distinction between "corrupted nature" and "perfected nature," the author(s) could not attempt to draw down a higher order of nature into the terrestrial sphere through distillation.

The Pseudo-Arnaldian authors also did not include Joachite apocalyptic elements within their works (as I shall discuss further in the next chapter). Because they did not mention Antichrist, the spiritual men, or the third state, they could not suggest either transmutatory or medical alchemy as a means to ameliorate the apocalyptic crisis. Furthermore, the genuine Arnald's prescription to heal the ills of Christian society in the last days comprised a set of standard spiritual and political reforms; he did not suggest that alchemy or medicine per se offered any remedy to the crisis. Only Rupescissa brings together quintessence medicine, the perfected nature of "human heaven," and the tribulations of the last days into a cohesive religio-alchemical system.

Despite these differences, Arnald's life and works perhaps bear the greatest similarity to those of Rupescissa. Judging from Rupescissa's citations of Arnald within his works and his ready acceptance of Arnald's apocalyptic predictions, Rupescissa clearly viewed his predecessor as both a great alchemist and a great exegete. Rupescissa, like Arnald, found within contemporary eschatology an imperative to search not only biblical Scripture but also the mysteries of the natural world for clues to the last days. Whether or not Rupescissa modeled himself consciously upon the outspoken physician, the parallels between Arnald and Rupescissa demonstrate a late-medieval current of apocalyptic thought wedded to naturalist investigation.

THE ARTS OF RAMÓN LLULL

Arnald of Vilanova had ties to another naturalist and religious reformer, his contemporary and compatriot Ramón Llull (c. 1232–1316), with whom he shared a great deal both intellectually and spiritually. Llull and Arnald were often viewed as similar thinkers by their contemporaries, and modern scholars have likewise tended to discuss the two men together.[60] The many writings attributed to Ramón Llull present some of the same challenges as the Arnaldian corpus. Llull's writings actually comprise two distinct bodies of work: those writings traceable to the genuine Llull and those by later writers who placed his authoritative name upon their texts. The historical Ramón Lull was a Majorcan Franciscan tertiary best known for his universal system of knowledge, the *ars combinatoria*, which was designed to aid the conversion of the infidel and the unification of the church.[61] The *ars* relies upon the symbolic use of alphabetical letters to represent God's "dignities," as well as other concepts linked to the dignities by analogy, such as the elementary qualities. Geometrical figures organize the letters into columns and wheels that may be rotated or otherwise manipulated in order to determine answers to a wide range of philosophical questions.[62] Building upon basic assumptions about the natural world shared by people of all faiths, the *ars* attempts to demonstrate Christian theological truths to Muslims and Jews by analogy rather than by biblical scripture. John of Rupescissa was so enamored of the *ars* that he devoted to its creator some especially lofty praise: "[Llull] has reached such a summit of philosophy that he has composed thirty-five books capable of uncovering all the sciences and resolving all problems, books so stupendous that the human spirit can hardly believe it unless it has experienced them with the intellect; and he has written many other books, around two hundred in all."[63]

Although Llull's focus was conversion, his interests touched upon a large number of topics, including theology, law, natural philosophy, and medicine, all of which he considered to be linked. His discussion of medical metaphors in the *Liber principiorum medicine* is of particular interest. Like Arnald, Llull envisioned a spiritual goal for naturalist knowledge.[64] He claimed that an understanding of the metaphorical signs of disease (by which he meant symptoms) led to a deeper comprehension not only of medicine but also of other forms of knowledge, including theology. Thus, he explained, the idea of complexionary equality provides a theoretical basis for understanding the equality of the Trinity; similarly, the qualitative properties of fire and pepper are a means to understand the incarnation of Christ.[65]

Llull's method is significant for two reasons: first, he offers analogies based on naturalist concepts in order to explain doctrines of theological significance. As I shall demonstrate in the next chapter, Rupescissa also uses analogies between nature and theology (although they operate in the reverse direction) to assert a similar unity of naturalist and religious knowledge. Second, the *ars* enlists nature in an attempt to convert the infidel and reclaim the Holy Land, goals that were not without eschatological import. Llull also explored questions of eschatology explicitly in his writings, such as in the *Liber contra Antichristum* (Book Against the Antichrist), the *De adventu Messiae* (On the Advent of the Messiah), and the *Quaestiones quas quaesivit quidam frater minor* (Questions Asked by a Certain Franciscan). But Llull was more cautious than either Arnald or Rupescissa: he refrained from predicting the exact date of the eschaton or claiming that Antichrist had already appeared. Nevertheless, Llull was deeply interested in apocalyptic themes, and these probably shaped his mission to reform the church and convert non-Christians. Llull may also have functioned as a model for John of Rupescissa in that he (unlike Bacon or Arnald) was not primarily a natural philosopher or a physician. He merely drew upon these intellectual resources to advance his main goal of converting unbelievers. Also, Llull noted in his writings that contemporaries called him *"phantasticus"* for his idealism and peculiar schemes. Rupescissa may have taken heart, since he was similarly named *fantasticus* during his trial before Clement VI in 1349.[66]

Although the genuine Ramón Llull wrote no works of alchemy, and was indeed skeptical of transmutation, an enormous corpus of alchemical texts bears his name, and some of these works are closely related to the works of Rupescissa.[67] The earliest of the Pseudo-Llullian texts—and one of the most interesting—is the early-fourteenth-century *Testamentum*. Little is known about the author of this Catalan work, who does not identify himself as Llull within the text. The figures and alphabets that appear in the work, however, are similar to those of the genuine Llull, and this apparently led readers to conflate the two by the end of the century. The *Testamentum* is especially important because of its views on the quintessence, as well as its praise for elixir as a singular cure for all ills. Likely influenced by both Roger Bacon and the Arnaldian corpus, the author of the *Testamentum* divided his work into theoretical and practical disciplines, and he deemed the alchemical elixir useful not only in the transmutation of metals but also in human medicine. According to the text, the "stone" or "medicine" is a single substance that transmutes metals into gold and silver, cures all disease, and even stimulates the growth of plants. The stone is so effective that there is no need for a diagnosis—it can cure any human illness.[68]

The author also mentions a "quintessence," which he argues is the substance of the entire universe, including the superlunary world and the bodies of the angels. A less pure form of the quintessence constitutes the sublunary world: the four terrestrial elements are actually corrupted versions of the quintessence that grow even more corrupt over time. When Christ judges the world at the end of time, everything made from these impure elements will burn. But there is still some hope. According to the author, a bit of the true and uncorrupted element remains inside matter, a "virgin" element that can withstand the fire of the last days. Such a material is of use to alchemists, since it can be used to synthesize a philosophers' stone.[69]

Pseudo-Llull hereby advances two important ideas about the quintessence. First, he views at least some version of the quintessence as heavenly, since it makes up the material of the celestial sphere; second, he locates a residue of the uncorrupted heavenly element within all matter. These doctrines may have had some influence on Rupescissa, although Rupescissa did not believe the four elements to be of the same substance as the quintessence. In fact, Rupescissa emphasizes that the quintessence is altogether outside the terrestrial elements, hence its power to overthrow corrupted nature. The *Testamentum*'s quintessence further differs from Rupescissa's in several other ways. It is a perfect balance of the four elements (which brings to mind the quintessence described in the thirteenth-century *De generatione stellarum*), its principle agent is not *caelum* (heaven) but *menstruum* (solvent, or sometimes menstrual blood) or *noster Mercurius* (our Mercury), and finally, it is a singular elixir that treats all illnesses.[70] A singular elixir is vastly different from Rupescissa's quintessence, which extracted properties from various medicinal materials and thus required specific diagnoses. Although the *Testamentum* shares some ideas about the quintessence with Rupescissa, it is far from a group of distilled medicines made from wine and other materials.

During the fourteenth century, unknown redactors integrated the alchemy of the *Testamentum* with that of Rupescissa's *De quinta essentia* and placed Llull's name upon the texts.[71] The most important of these Llullian-Rupescissan texts, the *Liber de secretis naturae seu de quinta essentia* (Book on the Secrets of Nature or the Quintessence), is actually a revision of Rupescissa's *De quinta essentia* that adds explicit discussion of transmutatory alchemy. The treatise, written in the second half of the fourteenth century, adopts most of Rupescissa's trademark ideas. First, the author praises the quintessence, a heavenly substance that can be found in all things, including plants, minerals, and animals (he includes tables indicating qualities and grades for different items within these groups). Second, the text discusses quintessences of wine, gold, and blood—all of which were central to Rupescissa's work. The author of *De secretis* also borrows the structure of *De quinta essentia*, dividing his book into two parts: the first book discusses the principles of the quintessence and its preparation; the second applies quintessence to specific ailments, including leprosy, paralysis, fevers, and other illnesses.[72] Quintessence can treat the illnesses of the body "from head to foot," the author claims, and, like Rupescissa, he argues that the quintessence can restore youth and prolong life up to its end as determined by God.[73] But Pseudo-Llull dispenses with Rupescissa's cautious avoidance of the word "alchemy"; instead, he unites quintessence medicine with metal transmutation explicitly.

The theory of the quintessence offered by *De secretis naturae* became widely diffused in the fifteenth and sixteenth centuries because of a proliferation of manuscripts and printed editions that far exceeded those of Rupescissa's *De quinta essentia*. A number of early modern physicians and alchemists pointed to Llull as the originator of quintessence alchemy, an innovation that they believed had been plagiarized by Rupescissa, since his career postdated that of the historical Llull. Texts attributed to Llull appear in several alchemical volumes edited by the Paracelsians or their sympathizers, including Guglielmo Gratarolo, Joannes Huernius, and Michael Toxites, the last of which attempted to demonstrate Paracelsus's dependence upon earlier adepts such as Llull.[74] Because the Llullian *De secretis* enjoyed such great influence, many ideas that are rightly Rupescissa's, including the distilled quintessence, have become attached to Llull.[75] A closer look at the genealogy of the quintessence, however, establishes Rupescissa's role in the creation of this significant set of ideas, and it points to the importance of untangling the roots of Rupescissa's innovation.

ALCHEMY AND THE APOCALYPTIC THREAT

Rupescissa's particular formulation of apocalyptic alchemy was unique, but it is clear that he was a part of a longstanding search for natural remedies to eschatological problems. Many of the crucial elements of Rupescissa's thought were already present in the intellectual currents of the later Middle Ages, but Rupescissa was the first to gather the strands of quintessence theory together into a cogent system of alchemy and apocalypticism. Rupescissa's significance is not only in his quintessence, however; it is also in the way in which he represents a broader effort to synthesize sources and draw connections between various disciplines. Strikingly, Rupescissa showed little regard for the institutional status of the disciplines he studied—university-sanctioned medicine and natural philosophy coexisted easily with transmutatory alchemy in his writings. When viewed alongside similar thinkers such as Bacon, Arnald, and Llull, such a method suggests that what modern historians perceive as disparate intellectual disciplines actually made up a coherent body of knowledge in the fourteenth century.

Furthermore, Rupescissa presented his theory of the alchemical quintessence as a purely Christian system, grounded in the structures and language of the Franciscan Spirituals and Joachite prophecy. He channeled history and theology together into a new genre of alchemy, one that was thoroughly Christian and thoroughly apocalyptic. The medieval apocalyptic tradition, as we have seen, furnished Rupescissa with a flexible hermeneutic system—a set of ideas adaptable to nearly any historical circumstance or body of knowledge. Apocalyptic prophecy gave Rupescissa a grand historical narrative with which to frame his alchemical studies. In alchemy, Rupescissa saw a solution to the financial woes of a battle with Antichrist, he saw cures for plagues and war wounds, and he saw elixirs for extending the lives of Franciscan Spiritual heroes. This system also inspired him to introduce apocalyptic language into alchemy itself, suggesting the possibility of drawing an otherworldly perfection into this world. The application of Joachite last things to naturalism made Rupescissa's idea of a quintessence plausible, and it made his alchemy, at its heart, apocalyptic.

Rupescissa's position within the context of writers such as Bacon, Arnald, and Llull points to an emerging current of thought in the Middle Ages that claimed human agency at the time of the apocalyptic crisis. As I have already shown, Rupescissa argued that human beings could use apocalyptic exegesis to survive the battle with Antichrist and populate the postapocalyptic millennium. He also contended that Christians could channel the perfection of heaven through alchemy and alter the course of human history. Other apocalypticists of the fourteenth century made different claims that nevertheless expressed similar assumptions: urgent times required extraordinary human action. Bacon advocated the study of natural science in the face of an attack by Antichrist, Arnald insisted that the church use its resources to predict the apocalyptic events to come, and Llull created his *ars combinatoria* in an effort to bolter Christian spiritual and eschatological claims during a critical period in Christendom. In each case, these thinkers saw the challenges of their day as a catalyst for human action. Yet Rupescissa went further than any of his predecessors in his assertions about the human potential to affect the events of the last days. Only Rupescissa argued that humankind could access the perfection of the heavenly sphere; only Rupescissa wrote that people could defeat Antichrist and survive into a thousand-year-long millennium. And only Rupescissa fully incorporated Christian apocalyptic elements into alchemy, allowing human beings to reenact the major events of salvation history through their craft, as I explore in the next chapter. Yet Rupescissa's ideas are better understood when they are placed within the strain of naturalist-apocalyptic thinking as it was developing during the Middle Ages.

Rupescissa's synthesis of Christianity and naturalism points to two important features of his alchemy, both closely tied to his eschatological concerns. The first is the practical function of alchemy against Antichrist. Although the problem of Antichrist is chiefly spiritual and theological, Rupescissa's answer to it is markedly materialist and practical. The second is the relationship between the alchemically perfected body and the Joachite third state of heaven on earth, which was to be initiated and populated by evangelical men, the very practitioners of quintessence medicine. The next two chapters will follow these two ideas as I address the problem of alchemical imagery in Rupescissa's texts. The core of my argument is that Rupescissa's alchemy cannot be understood without a serious appraisal of the eschatological themes and vocabulary—particularly the apocalyptic narrative of violence and death, redemption and resurrection—that provided its foundation.

Six

METAPHOR AND ALCHEMY

The vivid and complex imagery of alchemy has long interfered with scholars' ability to understand the science and its significance to medieval society. Enigmatic code names such as "green lion," "sun," "moon," and the crucified "Christ" are omnipresent in alchemical literature, even as such texts purport to reveal secret operations to readers. Early-twentieth-century scholars such as E. O. von Lippmann and Julius Ruska were among the first to study such code names (*Decknamen* in German). Their research, continued today by historians such as William R. Newman and Lawrence M. Principe, attempts to decipher the code names, in part by identifying them with chemical substances and practices.[1] Such scholarship has great value because it identifies the operative elements of medieval alchemy and establishes the continuity of alchemical traditions from the medieval to the early modern era. But the coded language in alchemical texts can give us more information about some of the foundational ideas behind alchemy.

Scholars have often assumed that the presence of coded religious imagery in alchemical literature is simply the product of a culture in which all aspects of life were steeped in religion. Historians who study alchemical code names thus tend to place little weight on which religious images appear in alchemical texts, and they have not attempted any sustained analysis of such images. The eminent historian of alchemy William R. Newman, for instance, has described religious language in alchemy as a "mystical flavor" that was not especially significant to medieval alchemy, which he argues was focused primarily on material processes rather than spirituality.[2] In contrast, I would suggest that such spiritual imagery is central to the way in which Rupescissa formed and communicated his alchemical theories.

METAPHOR AND ALCHEMY

Scholars of modern science have analyzed contemporary scientific metaphors, such as "greenhouse" gases, cell "factories," electron "orbits," and other concepts that have become standard ways of imagining our world in the twenty-first century. According to these scholars' arguments, the process of scientific metaphor not only accommodates language to "reality" but also shapes our understanding of reality by means of language. Scholars rightly hold that the selection and reception of such metaphors tell us a great deal about the societies that created them.[3] A similar project is possible with respect to the naturalism of the premodern period. By considering the specific images, tropes, and metaphors that appear in medieval alchemical texts, we can learn a great deal about authors' (and readers') assumptions about the human body and its place in the narrative of Christian history.

In this chapter, I examine the context and consequences of Rupescissa's code names, metaphors, and similes (which invoke heaven, Christ, the crucifixion, and blood, among other images) in order to deliver a fuller reading of their meaning. I evaluate Rupescissa's images by comparing them to those used in other alchemical and medical texts, in theological doctrine, in devotional literature, and in apocalyptic prophecy. As we have seen, Joachite apocalypticism provided a foundation for Rupescissa's eschatological views and for his alchemy. Belief in an imminent end drove his study of the natural world, since evangelical men who understood transmutatory and medical alchemy stood a better chance of defeating Antichrist. Here, I shall argue that Christian eschatological thought also offered images that constituted alchemy. Such imagery has three central functions in Rupescissa's work: first, it reveals his alchemy's theoretical basis—the eschatological model of corruption, purgation, and salvation that he used to understand natural processes. Second, it functions as a heuristic tool to make difficult-to-understand alchemical processes clear to readers by linking them to religious concepts such as heaven, Christ, or the resurrection. Third, it serves as a rhetorical device that imbued Rupescissa's naturalist program with the weighty significance of the apocalypse—the culmination of all of human history.

INTERPRETATIONS OF ALCHEMICAL IMAGERY

It would be impossible to treat here the wide range of code names that appear in medieval alchemical texts. I shall restrict myself to considering a few common varieties of code names and analogies prominent in the thirteenth and fourteenth centuries in order to put Rupescissa's vocabulary into context. Astronomical words are among the most common of code names, and they are an integral part of Rupescissa's lexicon. He names gold and silver the "sun" and "moon"; quintessence, "heaven"; copper, "Venus"; iron, "Mars"; quicksilver, "Mercury"; lead, "Saturn"; tin, "Jupiter"; and the healing herbs used in alchemical remedies, "the stars."[4] This sort of appellation was typical of alchemical texts circulating in the thirteenth and fourteenth centuries, including the works of Roger Bacon and Petrus Bonus of Ferrara; it also appears in the *Tabula smaragdina* (attributed to "Hermes Trismegistus"), the so-called bible of alchemy.

Biological terms such as "mother" and "father" were also common. For instance, in the *De mineralibus*, Albert the Great describes sulfur and mercury as the "mother" and "father" of transmuted metals; similarly, he compares the mineralizing power that developed stones (a process parallel to that of metals) to the formation of a fetus through seminal force.[5] The use of biological reproduction as a model for transmutation fits into a larger trend in the development of alchemy as a naturalist discipline. From the appearance of alchemy in the West until the late thirteenth century, alchemists used various analogies in their writings to portray the field as a new scientific discipline and to integrate it into the intellectual tradition of the university (a project that was unsuccessful, at least in institutional terms).[6] These analogies often relied upon the vocabulary of established fields of study, such as medicine or natural philosophy, which allowed authors to set forth alchemical operations in familiar and accepted terms.

In the early fourteenth century, a new type of code name appeared in alchemical texts. References to Christ, the crucifixion, and the resurrection pervade late-medieval alchemical literature, including Petrus Bonus of Ferrara's *Pretiosa margarita novella* and the writings attributed to Arnald of Vilanova and Ramón Llull. Barbara Obrist has argued that during this phase, such analogies no longer served to clarify alchemy with language based on university disciplines but cloaked it in highly convoluted metaphorical discourse. By this time, most alchemists considered their work incomprehensible through normal intellectual inquiry; instead, knowledge of alchemical operations required divine revelation.[7] This endowed alchemy with a supernatural element and established a new repertoire of terms with which to identify (or obscure) its operations.

METAPHOR AND ALCHEMY

During the nineteenth century, occultist scholars such as Mary Anne Atwood and Ethan Allen Hitchcock interpreted alchemy as concerned only with the interior spiritual development of the individual.[8] They viewed the descriptions of alchemical operations in early modern texts as allusions to the supramundane transformation of the practitioner. A later interpretation of alchemy by the Swiss psychologist Carl Jung similarly divorced the art from actual physical processes. Jung claimed that alchemy allowed for the projection of imagery, such as dragons, sexual reproduction, and the crucified Christ, from a vast "collective unconscious" onto the conscious mind. Both the occultist and Jungian models promote a "spiritual" understanding of alchemy that distinguishes it from chemical practice. Recent scholarship often unwittingly adopts this view by separating medieval alchemy into "esoteric" (interior and spiritual) and "exoteric" (exterior and material) strains.

It is therefore a healthy response for historians such as William R. Newman to remind us that alchemists were chiefly concerned with physical processes and material goals. Yet one need not draw a sharp contrast between materialism and spirituality in alchemy. Rupescissa's alchemy was above all material and operative, but it was nevertheless aimed at the solution of spiritual problems: Antichrist, the apocalypse, and the collective salvation of the Christian community. Moreover, Rupescissa built the theory of his alchemy upon spiritual concepts. I therefore argue that religious terms were more than mere flavor; for Rupescissa, they were an integral, organic part of his alchemy. They indicate its theoretical underpinnings, and they link it to the eschatological narrative of medieval prophecy.

IMAGES OF HEAVEN

In chapter 4, I discussed Rupescissa's identification of the quintessence, an alcohol-based remedy, with the Aristotelian fifth element, a celestial substance immune to generation and corruption. Rupescissa was likely the first to make such a connection between alcohol and the fifth element (also known as ether). Although earlier medical writers such as Taddeo Alderotti, Arnald of Vilanova, and Vitalis of Furno discussed medicinal alcohol as a powerful and unusual remedy, they did not identify it with a perfected, otherworldly element. For them, alcohol was merely *aqua vite* or *aqua ardens*.⁹ Rupescissa's use of the code name "heaven" to describe quintessence is new and extremely important.

In *De quinta essentia*, Rupescissa most commonly refers to the quintessence as *caelum nostrum* (our heaven) or *caelum humanum* (human heaven). He explains the expression:

> I shall demonstrate from the experience that I have had that [quintessence] confers incorruptibility and preserves from corruptibility, since if any bird or piece of flesh or fish is immersed in it, it will not be corrupted while it remains in it. Therefore, how much more will it preserve the animated and living flesh of our body from all corruption? This quintessence is the human heaven [*caelum humanum*] which the Highest created for the preservation of the four qualities of the human body, just as heaven is for the preservation of the whole universe. And I know for certain that Philosophers and medical doctors of our day do not know at all this quintessence and its truth and power. But with the help of God, I shall reveal to you below its magisterium and I [shall] teach you this secret thing, the quintessence: it is the human heaven.¹⁰

It is significant that Rupescissa uses the word "*caelum*," because in doing so he draws together two important aspects of the quintessence, as embodied in *caelum*'s dual meaning. *Caelum* refers to heaven in two different senses: first, the heaven or sky that contains the incorruptible bodies of the superlunary world and, second, the heaven of theological import, the place or state of blessedness after mortal life.

The first definition indicates Rupescissa's understanding of the quintessence with regard to the four elements of the body, the origin of its ability to prolong life. According to Rupescissa, the quintessence was equivalent to the fifth element of the celestial sphere, the substance that generated circular motion and perpetuated the existence of the superlunary world. The fifth element maintained its incorruptibility and immutability with respect to the four elements (earth, air, water, and fire), and it corrected cosmic imbalances by supplying a measure of any needed element. Medicinal quintessence similarly maintained a balanced complexion in the human body by adjusting the quantities of the four humors (blood, phlegm, yellow bile, and black bile).¹¹ A perfectly balanced complexion preserved the body in a state of youthful health; quintessence was thus a source of incorruptibility for human bodies, at least until their predetermined time of death.

An identification of quintessence with "heaven" helped Rupescissa to posit a parallel relationship between the astrological influence of the celestial bodies and the medicinal influence of alchemical remedies. He argued that the healing properties of gold, herbs, and other substances could be transferred to the human body by means of the quintessence. Quintessence therefore not only possessed its own medicinal capabilities but also served as a conduit for the "influences" of other materials, which were fittingly linked by code names to the sun, planets, and stars. Rupescissa writes that "just as the influx of the highest heaven affects the world and creates marvelous influence not only by itself but also by the virtue of the sun and the other stars, this heaven, the quintessence, wishes to be adorned with the marvelous sun [that is, gold] ... which is created in order to adorn our heaven and to augment the influence of the quintessence."[12] Rupescissa's use of the word "*caelum*" identifies the quintessence with Aristotle's fifth element and its ability to facilitate celestial influence while remaining immutable with respect to the four elements around it. "Heaven" thus operates as something more than a code name here: it reveals the basis of Rupescissa's quintessence in cosmological theory.

Scholars who have analyzed Rupescissa's "heaven" code name have generally recognized it as a reference to Aristotle's superlunary heaven and its role in cosmic balance.[13] But they have not probed Rupescissa's belief that the celestial world could penetrate the terrestrial world and fill it with immutable, quasi-immortal bodies. Why did Rupescissa think that the elements of two such incompatible realms could be combined? It is telling that Rupescissa chose to name the quintessence "heaven," which denotes not just the Aristotelian fifth element but also the quintessence's power to bring a key benefit of heaven—the perfected and immutable body—to human beings on earth before their death and resurrection.

Through the quintessence, humans could attain an absolute balance of humors, which would lead to the physical perfection and incorruptibility normally associated with immortal, postresurrection bodies.[14] This power was central to one of Rupescissa's main goals: helping readers achieve the extraordinary health and longevity needed to fight Antichrist in the last days. As in the writings of Roger Bacon, the effects of the alchemical elixir were bound to the perfection of the postresurrection body: static, balanced, and existing outside of time. When Rupescissa wrote that quintessence transmitted perfection to the things around it, it is this changeless, eternal perfection he had in mind. Quintessence, or "human heaven," was the fifth element, but it was also *heaven*—the state of eternity, immutability, and complete fulfillment. In this code name we can see Rupescissa's project to join two seemingly incongruous and perhaps even oppositional models of heaven: the Aristotelian natural-philosophical model and the Christian theological model.

Furthermore, *"caelum"* associates the quintessence, a product of the terrestrial world, with the sphere of the celestial. Rupescissa wrote that quintessence was a gift from God, created for the preservation of human bodies. It is found in nature, but it is not composed of any of the four natural elements. It is instead a divine gift, analogous in property and function to the realm of heaven. This indicates the quintessence's status as a piece of "perfected nature." It is also important to note that quintessence is not only "heaven" according to Rupescissa; it is "our" or "human" heaven. This use of the possessive reveals the critical role of human beings in securing perfected bodies through alchemy. The quintessence is a divine gift, but people must still prepare and refine it according to Rupescissa's instructions in order to receive its benefits. "Human heaven" also emphasizes the dichotomous nature of the quintessence, since it brings together two seemingly distinct, and even incompatible concepts—humanity and heaven—into one phrase. The phrase describes the quintessence's creation by the paired force of humanity and divinity, and its power to draw together the perfect superlunary world and the imperfect sublunary world. The quintessence functions as a bridge between the two spheres, allowing aspects of heaven to enter the world of humanity.

Rupescissa's association of quintessence with the Christian heaven is understandable in light of his adherence to the apocalyptic program of Joachim of Fiore. Joachim predicted that during the third and final state of history, the natural world of humanity and the supernatural world of divinity would meet. Humans would live in an age of the Holy Spirit, a perfected earthly utopia in which all would experience divine illumination and fully understand the mysteries expressed in holy Scripture. Rupescissa believed that this final age was near: his *Vade mecum* claimed that the appearance of Antichrist and his apocalyptic disasters could be expected in less than a decade; a Joachite earthly paradise would soon follow. Rupescissa's vision of alchemy thus parallels the Joachite view of the apocalyptic third state, in which distinctions between the heavenly and the earthly worlds collapsed. According to Rupescissa, the alchemical knowledge needed to secure "human heaven"—quasi-immortal, postresurrection bodies on earth—would be achieved in the context of the last days, as humans entered the third state of history and became filled with the Holy Spirit—also a "human heaven" on earth. Because Rupescissa's work was so permeated with this expectation of the millennial third state, I argue that the idea of the Joachite heaven on earth inspired Rupescissa to imagine the quintessence as a *caelum humanum*.

Rupescissa's use of "heaven" as a code name is clearly more meaningful than a simple case of "mystical flavor." Such language actually tells us a great deal about the formation of quintessence theory, and it indicates a deep interrelation of Rupescissa's prophetic and naturalist endeavors. The image of heaven in his writings reveals that he constructed his alchemy from the same eschatological narrative that underlies his prophecy. "Heaven" points to his goal of hastening the benefits (and, in some cases, the pain) normally experienced by human beings after death, judgment, and even resurrection so that they could be experienced within time and on earth. Through the alchemical agent, a timeless, otherworldly perfection digests and purifies the corruptible materials of the sublunary world. This heaven is strikingly similar to the heaven on earth of the third state, which arises from the corrupt preapocalyptic age after a similar cleansing and purging and which delivered the promise of heaven to humankind before the general resurrection and the end of time. The perfect structure of the universe and of the future millennial state suggested to Rupescissa the possibility of human perfection.

IMAGES OF CHRIST

With the appearance of alchemical works by (Pseudo-)Arnald of Vilanova and Petrus Bonus of Ferrara in the early fourteenth century, identifications of the philosophers' stone with Christ were becoming widespread.[15] Petrus Bonus's *Pretiosa margarita novella*, for instance, was among the first Latin works of alchemy to point out a parallel between the development of the philosophers' stone and the life of Christ. Petrus noted that during the alchemical process the stone

> appears as if dead; thereafter it requires fire, [until] the spirit of that body is extracted ... and the body becomes dust. After these things have been done, God returns to it its soul and spirit, and its weakness is washed away. It is

strengthened greatly, and after lightning it is improved, in the manner of man after the resurrection. ... On this account, this conception is similar to the conception by a virgin, which conceives without man: this is not able to be except by miracle, namely by divine grace.[16]

For Bonus, the stone bore all the chief aspects of Christ and his life: it was a union of divine and natural elements, the product of a miraculous birth, and subject to death and resurrection. Moreover, the stone had a soteriological function that further linked it to Christ's ability to redeem humanity.[17]

The various alchemical treatises attributed to Arnald of Vilanova made similar parallels between Christ and the alchemical medicines. In *De lapide philosophorum* (also known as *De secretis naturae*), the author organized his description of the alchemical operation with respect to the life of Christ, ending the work with a symbolic death and resurrection.[18] Other texts bearing Arnald's name extended this identification, describing the "scourges" of the metal mercury, in parallel to the instruments of Christ's Passion. Lynn Thorndike notes that the parallel between alchemy and Christ's life was so popularized by works attributed to Arnald that subsequent alchemists generally cited it as his invention.[19]

Rupescissa no doubt viewed Arnald as an authoritative author of both alchemical and prophetic works, and he included in his writings a number of quotations from Arnaldian writings. He drew upon the genuine Arnald's *De tempore adventus Antichristi* and *De mysterio cymbalorum ecclesiae* to develop his millennial dating system, and he found in Arnaldian alchemical sources a similarly attractive tool to serve his apocalyptic-naturalist agenda. In particular, he found in them a means to equate the philosophers' stone and its operations with key episodes in the life of Christ, allowing him to draw a parallel between the alchemical opus and Christian redemption. "Magister Arnaldus" is the only alchemical authority that Rupescissa quotes by name in the *Liber lucis*, and a majority of these passages identify the stone or its creation with Christ, his crucifixion, or his resurrection.[20] Rupescissa's choice of these Christocentric passages served two important purposes. First, Rupescissa used Arnald's alchemy to develop his own theory and practice. Second, Rupescissa decoded and expanded upon Arnald's code names in order to make alchemical ideas comprehensible to his readers. This latter method was parallel to the way in which his apocalyptic exegesis made the meaning of prophetic scripture clear to his audience in his prophetic writings.

Rupescissa's method in the *Liber lucis* is in fact very similar to that in his prophetic treatises. In his effort to build an alchemical theory and communicate it to his audience of "evangelical men," Rupescissa lists the names of alchemical writers such as Hermes, Alphidius, Geber, and Arnald of Vilanova. Rupescissa presents these sources as authoritative, even prophetic adepts of alchemy, "to whom this mystery has been revealed through revelations."[21] Rupescissa then subjects the figurative language of alchemy to an exegesis of sorts, in which he identifies code names with actual chemical substances and alchemical procedures. Such decoding of authoritative texts by an alchemical writer was a traditional means to impress readers and stake a place for oneself in the company of adepts.[22] But Rupescissa's exegetical process also resembles his approach in his spiritual treatises, in which he interprets the apocalyptic characters of the Bible and other prophetic texts as ciphers for the political and religious figures of his day.

In one passage, Rupescissa quotes from the "*Tractatus parabolicus*," an unidentified Arnaldian work that may be related to the *Rosarius philosophorum*:[23]

In the *Tractatus parabolicus de maiori edicto*, under the figure of the Gospel and the crucifixion of Jesus Christ, [Arnald] of Vilanova said these words: "Unless the grain falling into the ground dies, it remains only one, but if it dies, it brings forth much fruit" (John 12:24). Understanding the grain to be quicksilver mortified in the earth of saltpeter and Roman vitriol, and there mortified through sublimation by fire, it brings forth much fruit, because this is the great stone that all the philosophers sought.[24]

Rupescissa's interpretation of this quotation is important for two reasons. First, it demonstrates his method of alchemical exegesis. The structure of the passage (quotation of prophetic authority, followed by alchemical or apocalyptic interpretation) is common to both his alchemical treatises and his commentaries on biblical and prophetic sources. Moreover, the passages that Rupescissa selects for exegesis in both genres of his writing tend to be eschatological in content. Such parallel themes and structures in the body of Rupescissa's work further suggest the intellectual unity of his naturalist and spiritual endeavors.

Second, the passage has relevance for Rupescissa's theory of the alchemical opus. Arnald's words paraphrase John 12:24, in which Christ announces his imminent crucifixion and suggests that his death would stimulate new life, just as the death of the seed of grain led to the growth of new plants in the field. This link between death and rebirth resulted in part from observations in the Middle Ages that new forms of life often sprang from death and rot in the natural world.[25] Descriptions of creatures spontaneously generating from decaying flesh were commonplace in medieval naturalism: bees supposedly derived from decaying calves, beetles from horses, scorpions from crabs, and locusts from mules.[26] Rupescissa's quotation of Arnald suggests that he believed that death was essential for the generation of new and perfect forms in the metallic world. And indeed, in chapter 5 of the *Liber lucis*, Rupescissa cites the putrefaction of metal as a crucial stage in the creation of the philosophers' stone. He instructs readers to heat the potential philosophers' stone, turning it into blackened terrestrial matter:

continue the fire until you see the material to be blackened, because if it is not slowed down too much and if the fire is increased to it, when you have seen the material blackened, rejoice because you have the beginning of digestion. Then continue the fire until all other colors pass away and you will see the material whiten a little. Then increase the fire under it gradually and imperceptibly so that you see the material whiten more and more and, when you have seen the material to be completely whitened, it is perfected and then you have the stone perfected to white, which converts all metals and metallic bodies into perfect silver better than that which comes from the mines.[27]

The black matter was earthy, putrefied, and dead—and thus able to give rise to new forms by means of transmutation; without this passage through death, the philosophers' stone would remain sterile. Such a formulation appeared frequently in medieval alchemical treatises. The *Speculum alchimiae*, attributed to Arnald of Vilanova, warned that if the alchemist tried to proceed without putrefaction, his work would be in vain.[28] Pseudo-Arnald further theorized that the perfection of the stone resulted from its death and decay: "O blessed nature," he wrote, "blessed is your industry, because you make the imperfect to be perfect through the true putrefaction, which is dark and black."[29] The black putrefaction produced by alchemy was chaotic and unformed matter—some alchemists even labeled it prime matter—and it was parallel to the substance of the universe before creation.[30] The journey of the stone from death to putrefaction to rebirth was symbolized by its sequential color changes: in the *Liber lucis*, it turned from the terrestrial, dark "*nigredo*" to the snowy, celestial "*albedo*" of the moon, and finally to the bloody "*rubedo*" of the sun.[31] This color sequence represented the death, rebirth, and perfection of the stone, and its subsequent ability to "bring forth much fruit"—that is, the creation of gold and the prolongation of human life.

As Betty Jo Teeter Dobbs has pointed out, biblical authority could be claimed for the identification of Christ with the philosophers' stone. It relied upon the exegesis of two mentions of "stone" in the Old Testament. First, in Psalm 117, a stone that was initially rejected by builders becomes the cornerstone of a building; second, in Isaiah, the Lord is described as "stone of stumbling" for those who do not sanctify him.[32] References to these two passages appear in the New Testament, and in both cases "stone" is interpreted as a metaphor for Christ.[33] This symbolic connection was adopted by medieval alchemists, who consequently linked the philosophers' stone to Christ in their works.

Perhaps the most striking comparison of Christ and the philosophers' stone appears in Rupescissa's exegesis of another quotation from Arnald of Vilanova in the *Liber lucis*:

From this fourth operation the expression Milk of the Virgin derives, about which Master Arnald says: "Because it is proper that the son of man be exalted into the air from the earth by the cross," which literally means from the material digested in the third operation, in which mercury is placed in the bottom of the vessel for dissolution, because what ascends from there is pure and spiritual, and converted into powdery air and exalted in the cross of the head of the alembic just like Christ, as Master Arnald says.[34]

This passage makes an explicit identification of the philosophers' stone with Christ: it *is* pure and spiritual, and it is *just like* Christ. Rupescissa interpreted Arnald's remarks about Christ as a reference to the extraction of the Milk of the Virgin (an ingredient of the philosophers' stone and another example of Rupescissa's Christian code names) from the digested terrestrial matter left in the bottom of the alchemical vessel during a previous stage of synthesis. The extracted milk was distilled, during which it rose in droplets "like Christ" to the top of the "cross"; Rupescissa thus identified Christ with the substance of the philosophers' stone and the cross with the alembic (the apparatus used in distillation). According to his exegesis, the philosophers' stone was liberated from the earth and raised into the air by means of the alchemical operation. This process could be interpreted either literally (the element of earth is removed from the stone through calcination and the substance rises into the air through distillation) or metaphorically (the philosophers' stone transcends the bounds of the earth and moves toward the exalted realm of heaven through the mediation of the cross). As is true for Christ, the soteriological function of the philosophers' stone seems to derive from its crucifixion (and its death and putrefaction). It reaches heaven only once it has been put on the cross of the alembic.

Nor is the head of the alembic the only cross to be found in the making of the philosophers' stone. Rupescissa explains the construction of the alchemist's oven through an analogy of his own: the iron sheet that provides the foundation for the stove should be made "strong and round and having four arms in the manner of a cross."[35] Such language further connects the alchemist's instruments to those used in the crucifixion of Christ, an important point since the alchemist will soon bring about the "death" of the philosophers' stone by heating it on this cross-shaped sheet. This parallel to the passion provides a means to clarify the creation of the philosophers' stone in light of a familiar narrative of Christ's life and death. If alchemy could be shown to mirror the passion, and the elixirs to mirror Christ, then their curative and preservative functions could be made comprehensible to readers.

Rupescissa pairs passages from Arnald with his own insights in the *Liber lucis* in order to identify alchemical operations with the passion of Christ—the central drama of Christianity. Such an identification worked to assimilate the quintessence and the philosophers' stone to Christ's body and to dramatize the alchemist's operation as a reenactment of Christ's death and resurrection. Such use of the stone-Christ analogy appears not only in Rupescissa's alchemical treatises but also in his prophetic writings. In the *Liber secretorum eventuum*, he analyzes a passage from Daniel (2:34) and interprets the word "stone" as a reference to Christ:

The text says that "a stone detached from a mountain without the aid of hands strikes the statue's feet of iron and clay and reduces them to bits." It is evident that the clay feet, the part of the statue which is struck by the stone, are the times and the tyrants of the times of the Antichrist, and the end of the Roman Empire divided into ten toes.... Therefore, according to the principal sense, in the time of the ten toes—that is the ten kings—of clay the Roman Empire will be struck in the person of the Antichrist by the stone Christ and reduced to nothingness.[36]

According to this reading, the stone that will batter the imperial kingdom of the Antichrist during the last days is none other than Christ himself. This interpretation emphasizes Christ's role as a weapon against the Antichrist's army during the end-time battle. Rupescissa's exegesis falls squarely within the traditional interpretation of Christ as a stone of building or stumbling, as expressed in the New Testament.[37] Rupescissa makes a similar analogy in his *Vade mecum* (written seven years later), but in this version he explicitly connects the stone-Christ to the dawning of the Joachite third state and the appearance of the angelic reformist pope:

This man [the Repairer] is after Christ in the beginning of the third state of the earth, the stone torn away from the mountain without any hands, which will strike down the Babylonian statue in its clay feet and break it up, and through the stone Christ and his law it will complete the whole earth, converting all infidels to the lord, except in the corners of the world, where Gog and Magog will remain.[38]

In this case, Rupescissa identifies the Repairer as the new Christ who destroys the stone statue, which Rupescissa reads as a representation of Antichrist and his kingdom. The stone is again a weapon against Antichrist, but it is also now the initiator of the Joachite third state. Rupescissa's interpretation has developed to include not only the battle against Antichrist but also the dawning new age of peace and illumination.

It is probable that these passages also play upon the dual meaning that "stone" had for Rupescissa: stone was an ordinary rock, used to build or destroy, but it was also the philosophers' stone, the alchemical elixir used to perfect metals and bodies. In light of this, the above passages point to the apocalyptic function of the philosophers' stone, since Rupescissa held that the stone, like the Repairer's appearance and Christ's return, would be possible as the end of time approached. The philosophers' stone could thus serve as a weapon against Antichrist and a tool of the spiritual men at the beginning of the third state. Rupescissa draws together Christ, the stone, and the philosophers' stone as the multivalent means of striking down the forces of Antichrist and fortifying true Christians with spiritual and physical remedies.

Rupescissa's focus on the problem of an imminent end has several consequences for his alchemy. While other alchemists included discussion of the effect of alchemy on the alchemist's soul, such interior transformation appears to interest Rupescissa not at all. According to Petrus Bonus's *Pretiosa margarita novella*, for instance, the successful alchemist received a special revelation from God that was conveyed only to spiritual adepts and that assisted in the alchemical process. Petrus Bonus moreover claimed that the biblical prophets, including Moses, Daniel, Solomon, and John, possessed knowledge of the alchemical art through divine revelation. According to Bonus, these prophets were interested in alchemy for the insight it provided into religious matters, as well as its purifying effects upon the soul. Because of this, the ancient alchemists were able to intuit the coming incarnation, as well as the doctrine of the Trinity.[39] Much has been made by scholars about the meaning of this "esoteric" form of alchemy that catalyzes a spiritual transformation of the human interior. Eric John Holmyard has described esoteric alchemy as a "devotional system where the mundane transmutation of metals [becomes] merely symbolic of the transformation of the sinful man into a perfect being through prayer and submission to the will of God."[40] Interest in the spiritual benefits of "esoteric" alchemy (as opposed to "exoteric" alchemy, or metal transmutation, though the two forms were often mixed) was to become important during the early modern period, and the theme is developed by writers such as Jacob Boehme and Robert Fludd. A reference to this form of alchemy even appears in Martin Luther's *Table Talk*.[41]

As I noted above, historians of science have in recent years tended to emphasize the practical, or "exoteric," dimension of alchemy over its spiritual, or "esoteric," function. My evidence supports this interpretation: Rupescissa gives every indication that he intends his readers to engage in actual chemical practices rather than in mere contemplation. Yet Rupescissa's chief aim in creating the quintessence was doubtless spiritual. In his works, the operative, material function of alchemy takes on a religious weight: it provides for the physical needs of a church on the brink of an unprecedented spiritual crisis. In light of this, it is possible to emphasize the operative, practical nature of Rupescissa's alchemy without losing sight of its foundation in religious concerns.

IMAGES OF BLOOD

A number of analogies involving blood appear in both the *Liber lucis* and *De quinta essentia*. These images appear to link the philosophers' stone and the quintessence to the healing power of blood, both human and divine. But blood was no mere symbol in Rupescissa's alchemy. He also includes a recipe for quintessence that recommends the use of real human blood as a medicinal ingredient. The assumptions that Rupescissa makes about the soteriological and therapeutic properties of blood seem to derive from a variety of naturalist and religious sources: Pseudo-Avicennian elixir alchemy, Galenic humoral theory, and Christian devotional literature, as well as biblical and hagiographical accounts of violence and self-mortification. I shall offer some suggestions about how these various traditions work together to make up Rupescissa's complicated blood-based alchemy.

An important analogy between blood and alchemical materials occurs in *De quinta essentia*, in a chapter devoted to the derivation of quintessence of antimony.[42] Rupescissa writes that once the antimony is combined with vinegar and distilled, the reader should notice a "stupendous miracle": in the distillate "thousands of little veins of liquors of the blessed mineral descend in the form of little red droplets just as if it were rightly blood." This liquid is a "manifest miracle" worth more than all the treasures of the world because it "carries away the pain of all wounds and heals marvelously. Its virtue is incorruptible and miraculous and useful above all."[43]

A similar analogy appears in the *Liber lucis*. According to Rupescissa, the creation of the red philosophers' stone involves several bloodlike substances at different stages of its production. Among them is a "Tincture of Redness of the spirit of vitriol," which Rupescissa claims is the animating force of the philosophers' stone. He supports this claim with a quotation from "certain philosophers" stating that Milk of the Virgin is "alive with its own blood." Rupescissa follows the quotation with his exegetical remarks: the milk is "alive (*animal*) because it comes into existence through sublimation and because it has a soul (*anima*) of blood, that is, the Tincture of Redness with the aforementioned spirit of Roman Vitriol."[44] According to Rupescissa's interpretation, the philosophers' stone derives its color and character from its soul of blood, the Tincture of Redness. This was also the source of its force or animation.

The color red, which was suggestive of blood, also plays a role in the final stage of the making of the philosophers' stone. Using Arnald of Vilanova's writings as a guide, Rupescissa details the method for "finishing" the white philosophers' stone to red, its final stage of preparation before it could turn base metals to gold. To reach completion, the stone had to be extracted from the alchemical vessel, where it was enclosed, according to

Arnald, "just like Christ inside the sepulcher."[45] Rupescissa interpreted Arnald's remarks to be a reference to the digestion of the aforementioned Tincture of Redness by the heat of a flame so that the internal redness of the philosophers' stone could become externally visible, turning the stone from white to red. After this had been accomplished, the stone would, as Rupescissa explained, "ascend from the sepulcher of the Most Excellent King, shining and glorious, resuscitated from the dead and wearing a red diadem, just as Master Arnald has attested."[46]

Rupescissa's exegesis of Arnald's remarks claims a figurative "life" for the philosophers' stone. It is enclosed in a sepulcher; this implies that it has in some sense "died" and been entombed by the alchemist's equipment. It is subsequently "resuscitated" by the alchemist and emerges from the sepulcher. As it ascends, the stone wears a red diadem or crown. This analogy makes an explicit connection between the philosophers' stone and Christ, both of whom were killed, buried, and revivified. Similarities between the stone and Christ suggest one of the central messages of Rupescissa's alchemy. The creation of the stone through alchemy reenacts the narrative of the passion; hence, the stone may reenact the curative and salvific powers of Christ. The two redeemers are linked by their parallel experiences of death on the cross and subsequent resurrection and ascension.

Rupescissa's mention of the red diadem in the passage is also significant. The Tincture of Redness, or blood, of the stone is linked by color to a red crown on the head of the resurrected Christ. The red color suggests a connection to Christ's blood, and it further highlights the significance of blood to the stone's soteriological function. Moreover, because the internal redness of the stone becomes external through the operation, Rupescissa's alchemy is capable of making the invisible bloody nature of the philosophers' stone—the key to its function, and to the function of Christ's passion—visible. Like Christ, the stone is sealed inside the tomb of the alchemical vessel. The philosophers' stone rises and departs from the tomb, crowned and revivified, with the result that metallic bodies are redeemed from their base natures and perfected to the incorruptible nature of gold. According to Rupescissa's analysis, the stone is animated by a soul of blood that becomes visible through the alchemical operation, making clear the centrality of blood to the stone's and to Christ's salvific power.

Rupescissa's interest in alchemical medicine that relied upon the manipulation of blood was not without precedent. Roger Bacon had already written in his commentary on the Pseudo-Aristotelian *Secretum secretorum* that human blood was the ideal "stone," or base for an alchemical operation.[47] Based on his reading of Pseudo-Avicenna's *De anima in arte alkimia*, an Andalusian work of the twelfth century that had a great influence on Western elixir alchemy, Bacon interpreted Pseudo-Aristotle's "animal stone" as a code name for human blood.[48] He recommended that readers separate blood into the four elements through distillation, recombine the elements in "equal" proportions, and then mix the resulting substance with metallic calxes and mercury to create a philosophers' stone. Much as does Rupescissa's *De quinta essentia*, the text of *In arte alkimia* details the preparation of elixirs from various organic materials, in particular, hair, eggs, and human blood. For the author, blood was the preferred "stone" because it was the most noble and vital of spiritual substances, and because it was a composite of the four humors.[49] Bacon likewise preferred blood over all other substances because he believed that it was the closest thing to prime matter that could be attained.

A work attributed to Arnald of Vilanova, *De sanguine humano*, taught a similar extraction of healing materials from human blood.[50] This tract recommended a blood-based elixir because of the humor's special composite properties: the four qualities within it could be separated through distillation and applied to medicinal purposes.[51] The use of blood to prepare elixirs of health was thus fairly established in alchemical texts before it found expression in Rupescissa's *De quinta essentia*. It was to become a widespread formulation in part through his influence, if one can gauge the popularity of the treatment from the blood-based "quintessence" recipes that Piero Camporesi has collected from sixteenth- and seventeenth-century medical texts.[52] The descriptions of these medicines follow closely the language of Rupescissa's quintessence alchemy, and they repeat assumptions about blood's reparative effects as expressed in the *Liber lucis* and *De quinta essentia*.

But Rupescissa's use of blood goes beyond the Baconian and Arnaldian focus on the separation or recombination of blood through distillation in order to produce an elixir. He attributes special powers to blood, which he describes in his alchemy as a catalyst for growth or repair. This suggests that other blood traditions contributed to the role of blood in Rupescissa's work.[53] During the Middle Ages, Christ's blood became an increasingly powerful symbol in visual art; miracles involving blood (including stigmata and bleeding hearts) proliferated during this same time, as did devotion to supposed relics of Christ's blood and bleeding hosts.[54] As a number of scholars have observed, blood was often represented in medieval devotional literature as a catalyst for spiritual development or conversion.[55] In one fifteenth-century Carthusian miscellany, for instance, a vividly colored devotional image shows the tree of life sprouting from Christ's heart and irrigated by numerous droplets of his blood. At the tops of the sprouts are buds that "flower" into miniature portraits of the Virgin Mary and St. John, their hands clasped in prayer and their eyes on the bleeding Christ at the center of the illustration (see figure 2). Blood rains from Christ's body, providing the water and nutrients needed for the tree to grow and bloom. The image points to Christ's sacrifice and its parallel effect on the reader. Christ's blood brings about a religious transformation in Christian devotees that culminates in "blossoming" through meditation. This process is expressed by the floral images of Mary and John, which result from the growth stimulated by Christ's blood. The scholar Marlene Villalobos Hennessey writes that the tree of life in this image, which interprets the text of the scroll in the illustration, embodies the "fruits" of the passion and restores the reader to an almost prelapsarian, paradisical state of fulfillment.[56]

A German illustration of the late fifteenth century depicts a similar scene in the center of a field of poppies (see figure 3). Jeffrey Hamburger includes this image in his analysis of a group of nuns at the convent of St. Walburg in southern Germany, who created such paintings as aids to personal devotion.[57] Several of the images Hamburger discusses focus on the figure of the crucifixion, and one painting in particular portrays an oversized heart hanging from the cross. In this painting, vibrant, red drops of blood decorate nearly every part of Christ's body. As in the earlier Carthusian illustration, copious amounts of blood pour down onto vegetation below the cross. Brightly colored poppies gather in a grassy field, and they are densely clustered at the foot of a ladder reaching toward Christ's body on the cross.[58] The message is clear: Christ's blood nourishes and vivifies the poppies, as well as the Christian adherent who approaches Christ.

In these images, the stalks and blooms of plants are showered by blood, which functions as a fertilizer, facilitating growth or "flowering." Christ's blood is shown to be generative and fecund, promoting spiritual as well as organic development.[59] This idea appears in medieval written texts as well. The *Vitis mystica*, a text that has been attributed variously to Bernard of Clairvaux or to Bonaventure, advised readers to immerse themselves in the blood of the lamb: "Rest in it in order to become warm; once warm, become softened; once softened; let flow truly a fountain of tears."[60] As in the devotional image of the tree of life, the author connects Christ's blood

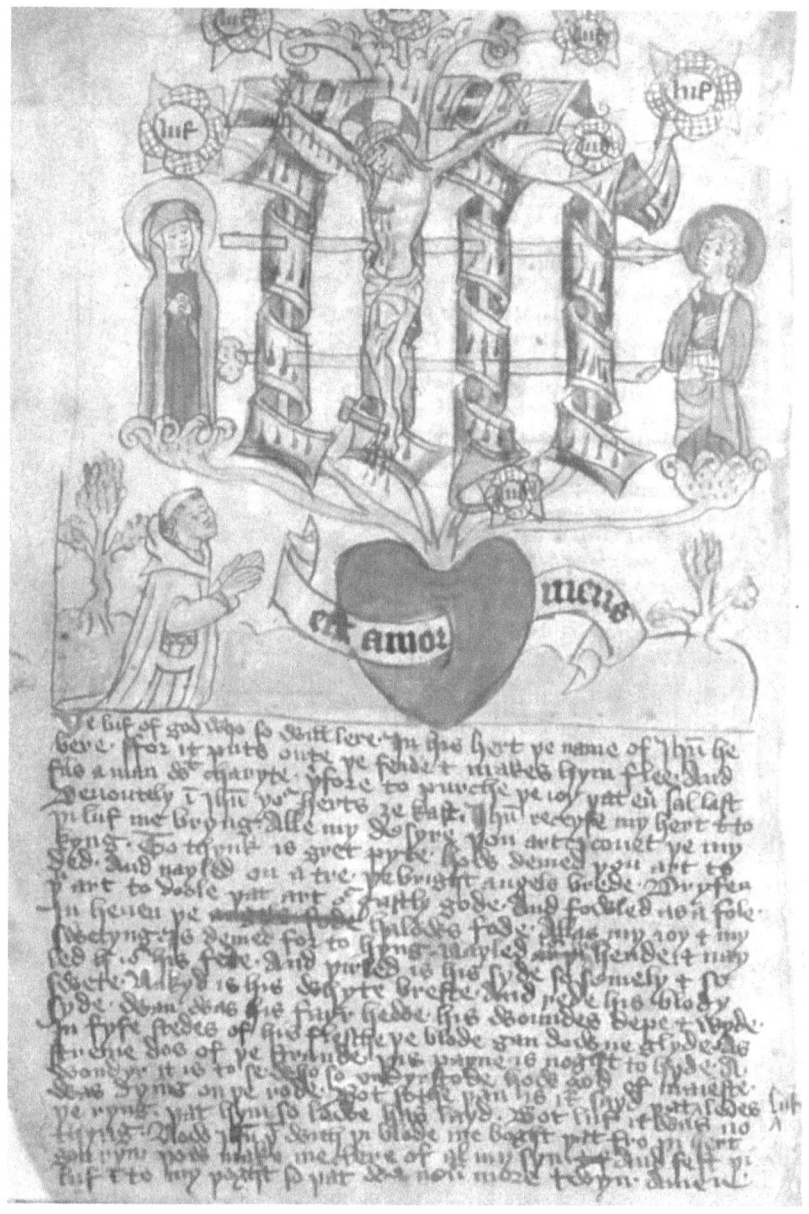

Figure 2. Tree of life watered by Christ's blood. *Source:* London, B.L. Additional MS 37049, fol. 36v.

Figure 3. Christ's blood fertilizes poppies beneath the crucifix. *Source:* Eichstätt, St. Walburg (c.1500). Photograph by Jeffrey Hamburger. Reproduced with the permission of the Abtei St. Walburg.

to a spiritual transformation, in this case a dramatic softening of the heart and an outpouring of grief and compunction. Similarly, liturgies associated with blood cults and defenses of blood relics in fifteenth-century Germany were rife with images of blood as vegetative, floral, and fecund. A number of miracle-host stories also associated blood with growth and fertility, including tales of laywomen who stole the host in order to enrich their businesses or romantic relationships through its generative powers.[61]

Devotional images of Christ's blood as transformative or generative are useful for understanding the role of blood in the *Liber lucis* and *De quinta essentia*. While Rupescissa emphasized a physical rather than a spiritual transformation, blood or bloodlike substances (generated through the creation of the philosophers' stone and the quintessence) served as alchemical fertilizers or catalysts for development. In *De quinta essentia*, Rupescissa recommended human blood as a base for a particularly effective quintessence:

> If you heard what I revealed to you above from the greatest secrets when I said that quintessence is in everything and remains uncorrupted, the greatest [secret] will be when I teach you to extract [quintessence] from human blood and from the flesh of all animals and from eggs and similar things. Since human blood is the perfect work of nature in us so much that it augments what has been lost, it is certain that nature perfects and has perfected this quintessence so that without any preparation it transforms blood from the veins immediately into flesh. And this extraordinary quintessence is the greatest thing of nature to be had, because in it is the marvelous virtue of our starry heaven, and it performs divine miracles to the cure of nature, just as I will teach below. Therefore, receive from the barber-surgeons blood taken from all sanguine and choleric young men, so that you obtain it from those who enjoy good wine, and the watery or phlegmy [superfluous materials] are drained out.[62]

In this passage, we see an example of blood generating and facilitating organic growth. According to Rupescissa, blood is transformed instantly into pristine and healthy flesh by means of the quintessence. He attributes to it the "marvelous virtue of our starry heaven," placing it within the tier of "perfected nature," which has the power to purge corruption. Blood treated with alchemy thus facilitates the growth and repair of the physical body in order to achieve the quasi immortality of a balanced complexion. As in the tree-of-life and crucifixion images, blood is fecund and catalytic. There is, however, one important difference in the way blood operates in devotional imagery and in Rupescissa's alchemy. Rupescissa's writings emphasize blood's ability to return the body to a previously youthful and healthy state and to maintain that state in the future; this is not the "flowering" of the body into something altogether new. Rupescissa's interest in blood's catalysis of growth thus seems to be focused on its preservative, and not its transformative function.

The curative and generative capacities of blood as expressed in Rupescissa's writings and in Christian devotional imagery are rooted in ancient medical theory. According to Galenic medicine, blood had a special place among the four humors. The blood found in the veins was thought to be primarily blood, but also to contain smaller portions of the other three humors, which could be observed when a container of drawn blood was allowed to separate into its constitutive parts.[63] In the Galenic schema, blood was the base material of the human body. Ingested food was carried to the liver, where it was transformed into blood by concoction, or "cooking." It then served as the foundation for the other bodily fluids, such as breast milk and sperm, which were manufactured through further internal heating, coagulation, and distillation. Blood was also responsible for the nourishment of the womb, which was thought to be connected to the breasts and the liver by a *vena kiveris*, or "female vein," that ensured the production and transport of milk and menstrual blood to the fetus. Blood was also thought to develop into the bodily tissues and bones, as well as to produce the other three bodily fluids during its own manufacture.[64] Thus it is perhaps understandable that one finds blood—with its ability to generate physical growth and to transmute itself into the various parts of the body—at the center of the philosophers' stone and quintessence, the alchemical agents capable of transmuting one substance into another.

It is tempting to categorize all of these descriptions of blood and wine in Rupescissa's alchemy as suggestive of the Eucharist and its healing effect on recipients. Caroline Walker Bynum has noted that recent scholarship tends to interpret any mention of blood and body, eating and drinking, or bread and wine in medieval literature as a eucharistic reference. Yet Rupescissa does not use any eucharistic language to characterize blood in his alchemy. Bynum demonstrates that a non-eucharistic blood tradition also existed, and she suggests that blood has accusatory as well as consoling and generative aspects.[65] Blood was a symbol not only of eternal life but also of violation, torture, and blame. Images of the blood and wounds of Christ, such as those discussed by Hennessey and Hamburger, demonstrate the generative and reconciliative aspects of Christ's blood—but they also accuse his killers (who included all of humanity) and stimulate guilt among believers.[66]

Indeed, human blood was for Rupescissa both a positive symbol of regeneration and an ambiguous pollutant that connoted violence. Such ambiguity was mirrored in the contemporary stigma against surgeons (who had frequent contact with blood) and in prohibitions against the shedding of blood by clergy.[67] Despite—or perhaps because of—this ambiguity, the release of blood from the body was generally considered to be purgative, as is evident in medieval writings on the physical benefits of bloodletting and menstruation, which were thought to clear impure matter from the body.[68] Such benefits were well known to Rupescissa, who emphasized the cleansing release of blood not only in his alchemy but also in his description of the violent tribulations of Antichrist.

In the *Vade mecum*, Rupescissa depicted the downfall of members of various social and religious groups, detailing their execution by Antichrist's forces. The extravagant outpourings of blood from slain secular lords and clergy served not only as punishments for their sins but also as a purging of sorts for the entire community.[69] After the release of the blood of the sinners, the remainder of society would start afresh, embracing the *vita apostolica* and the new millennial regime. Rupescissa wrote: "Through the wounds [inflicted by the tribulations], the greatest part of the present perverse generation will be killed, so that the world will be renovated and the arrogant hardened [Christians] removed from it, so that the truth be brought back."[70] The spilling of blood was for Rupescissa violent and horrible, but it also had a medicinal function. Much like a bloodletting, the slaying of the reprobates worked to clear impure or corrupt matter from the body of society, and its effect was ultimately a healthful one for the ailing patient.

Rupescissa's prescription for the treatment of the body of society to some extent follows a medical model: impure and corrupt matter (the sinful or infidel) must be cast out for the health of the whole. This indicates that Rupescissa conceived of the body of society as a macrocosm of the human body (although he never made an explicit equation of the two). Following humoral theory, the individual body could be purified and maintained through the regulation of its various parts by addition and subtraction. At first glance, Rupescissa's recommendation for Christendom seems to be based wholly upon the same principle. In his book on Arnald of Vilanova's medicine, Joseph Ziegler analyzes Arnald's similar recommendations for social and religious reform at the turn of the fourteenth century.[71] Ziegler views Arnald's approach to spiritual reform as a natural extension of his role as a physician: he applied assumptions about bodily composition and balance to the larger body of society in order to treat its spiritual ills. My interpretation of the relationship between medical and

{125}

reformist activity in Rupescissa's thought differs from Ziegler's analysis of Arnald. Rupescissa's approach to reform is not a simple application of humoral theory to social and religious problems. His spiritual treatises demonstrate a hearty interest in the wholesale destruction of members of the body of society, including Muslims and degenerate Christians, rather than in their incorporation into and continued harmony with the larger group.[72] Strictly following Rupescissa's humoral model, which emphasized balance and stasis, such radical destruction of portions of the body would be a deadly treatment for the sick patient.

Rupescissa's prescriptions thus appear to be influenced not only by medical models of balance but also by long-standing traditions of pious self-mortification and bodily purgation. Accounts of the radical removal or mutilation of body parts that engendered sin were prominent in early Christian texts. Peter Brown recounts a dramatic self-mutilation described in the apocryphal Acts of John, in which a young man castrates himself with a sickle, announcing, "There you have the pattern and cause of all of this!"[73] Brown notes several incidents of actual or attempted castration in the early church: Origen, for instance, was rumored to have had himself castrated.[74] There was also biblical foundation for such piety. Matthew 5:29 warns that "if your right hand causes you to sin, cut it off and cast it from you, for it is more profitable for you that one of your members perish than for your whole body to be cast into hell."[75] Self-mortification according to this ideal has often been cited by scholars as particularly characteristic of fourteenth-century piety. Richard Kieckhefer has identified a number of late-medieval saints who removed or scourged parts of their bodies in order to avoid sin and encourage spiritual progress. Catherine of Siena shaved off her hair in order to remove "the cause of her grievous sin" of vanity; according to Kieckhefer, such behavior was intended both to atone for past misdeeds and to keep the body from further sin.[76]

The impetus to remove the part of the body that caused sin was probably an element in Rupescissa's recommendations for societal reform in the *Vade mecum*.[77] The destruction of Muslims or sinful Christians is not unlike the shaving of a head or the removal of a right hand: the health of the body necessitates the complete removal of its offending members. Rupescissa's recipe for the recovery of Christian society appears to be influenced both by the medical theory of purgation, which aimed for health through the harmonious interaction of parts, and by traditions of religious piety that demanded a more radical separation of the harmful part from the whole.

Rupescissa's blood imagery in *De quinta essentia* and the *Liber lucis* is therefore similar to his use of images of heaven and Christ to explain alchemical materials. Such names and analogies reveal a complicated blend of ideas derived from both naturalist and spiritual traditions. Like Rupescissa's other figurative terms, "blood" functioned as much more than a simple code name. As we have seen, in some passages he makes a prolonged comparison of the quintessence or the stone's chemical states to the blood of Christ. In Rupescissa's discussion of the quintessence of blood, however, he makes clear that the substance is to be created from actual human blood. The writings of Roger Bacon and (Pseudo-)Arnald of Vilanova on the distillation of blood demonstrate that alchemical authors occasionally used "blood" not only as a code name but also to describe the presence of actual human or animal blood in alchemical medicines. For Rupescissa, blood has a multilayered symbolic and literal meaning; it is not a simple substitution name for an inorganic chemical substance.

CONCLUSION

While historians tend to treat prophecy or theology and natural philosophy as separate categories (leading to the bifurcation of scholarship on Rupescissa into those studies devoted to the history of science and those devoted to the history of religion), Rupescissa did not distinguish between the various modes of thought. It is likely that some of the boundaries that scholars have drawn between religious and naturalist approaches in the Middle Ages simply did not exist for Rupescissa and his contemporaries, who had a much more fluid interpretation of what are now perceived to be distinct disciplines. The code names and analogies that Rupescissa chose to describe the alchemical elixirs—heaven, blood, and the death and resurrection of Christ—reveal the amalgam of theology, prophecy, and naturalism that informed his work. His language played upon the similarities between supernatural elements and the behavior of the alchemical medicines. The alchemical medicines were made from terrestrial products, but their transcendent properties and functions more closely resembled the celestial sphere of perfected nature. Such identifications seem to draw together the geography of the earthly and the celestial worlds, erasing distinctions between imperfect terrestrial nature and perfected heavenly nature. Because of this, Rupescissa's alchemical medicines could be both natural, derived from ordinary earthly materials, and otherworldly, behaving in a manner parallel to heaven.

Furthermore, such language identifies the natural processes of alchemy with the overarching narrative of salvation history. The dichotomous qualities of the alchemical medicines, along with their roles in attaining perfection for human and metallic bodies, closely associate them with Christ, the incarnate God, who similarly straddled the spheres of earth and heaven. This dichotomy within Rupescissa's alchemy indicates that a religious model was fundamental to his theory of the alchemical medicines. Furthermore, the term "human heaven" has a dual meaning. On the one hand, it is the substance that functions in the human body as Aristotle's fifth element functions in the sky, balancing the four elements in perpetuity. On the other hand, it is the material through which a form of the Christian heaven—perfect and timeless bodies untouched by corruption—is achievable on earth through a Christlike savior in the time of Christendom's greatest need, during the reign of Antichrist.

Rupescissa's preoccupation with the apocalypse no doubt guided his selection of code names with which to describe alchemical materials. Although Rupescissa could have chosen analogies from other disciplines, his writings overflow with religious imagery. Rupescissa's apocalyptic alchemy also points to the need to take seriously the language that alchemical authors used to describe their subjects. Rather than dismissing religious language as without particular meaning except inasmuch as Christian authors expressed themselves in Christian terms, scholars should consider such images more carefully. The coupling of ideas that prompted Rupescissa to imagine the quintessence as a "human heaven," for instance, had enormous consequences for the development of alchemical medicine, which became increasingly concerned with distillation technology and quintessence remedies in the centuries that followed. It is thus important to examine the images and terms in his work, and to understand what role they played in the formation and development of his alchemical ideas. For him, the alchemical medicines were important because of their relationship to the narrative of Christian history. As I shall discuss further in the next chapter, such imagery suggests that the reenactment of Christian history through alchemy could lead to the return of Christ and the defeat of Antichrist in a world on the brink of a new age.

 Seven

THE END OF NATURE

As we have seen, Rupescissa used code names and analogies based upon religious ideas in order to identify alchemical materials and practices: the quintessence was heaven, the elixirs were Christ or his blood, the alchemical apparatus was an instrument of the passion. This chapter will attempt to make clear why such language was so important. Rupescissa applied eschatological concepts to alchemy not only to understand the workings of the natural world and to communicate them to his readers but also to assert a place for nature in the teleology of salvation history. Rupescissa argued that divinity was embedded providentially in nature. Humans could discover prophetic clues within the natural world that warned of the *what* and *when* of the apocalypse. Furthermore, because quintessence provided a conduit for the perfected sphere of the heavens to enter the corrupted sphere of the earth, humans could harness a higher level of celestial nature and intervene in the last events of human history.

In this chapter, I shall focus on two different areas in which Rupescissa treats the participation of the natural world in divine history. The first section analyzes Rupescissa's astrological theory as a means to predict the details of the apocalypse. Rupescissa imagined nature as a text brimming with prophetic clues and linked by concordances to biblical scripture and apocryphal revelations. This identification of nature with scripture indicates his view of the created world as suffused with eschatological import. Second, I shall evaluate Rupescissa's contribution to a debate in alchemical literature about the superiority of natural products to human-made ones. Rupescissa's position reveals the way in which he believed humans could use alchemy to raise corrupted nature to a higher level of perfection, thereby performing through artifice medical cures that were beyond the ordinary course of nature. In both of these cases, the natural world is "apocalyptic"—revealing the transcendent meaning of human and eschatological history. Nature furnishes the means for humans to predict, prepare for, and engage with the events of the end.

THE NATURAL AS SCRIPTURE

Like many naturalist thinkers of the Middle Ages, Rupescissa envisioned a God as a craftsman who imprinted his will upon the human body and the cosmos through the tools of nature. Natural processes were a means for God to act in the world, and even quotidian natural events reflected God's plans for humanity. Rupescissa explained this phenomenon in *De quinta essentia*:

I know that all of the stars in the sky have their extraordinary influence by the will and the ordination of God, and that each star has its property over a certain and determined thing, just as the star of the pole has influence over adamant and iron, the state of the moon has influence over the water of the sea, the sun over gold, the moon over silver, the images of men in the sky over human bodies, the image of the ram in the sky over the rams of the earth and the head of man. And just as the carpenter who creates chests with the aid of an axe is still the creator of chests, neither does God govern the world any less. He gave an influx to the stars so that they would flow into things just as and however he wished, and not otherwise.[1]

Nature was, according to Rupescissa, a tool in the hand of God, analogous to the axe used by a carpenter to produce his wares.[2] The operations of celestial bodies were thus primarily natural expressions of the creator's will to affect the world. Rupescissa's description of the zodiacal sign of Aries is likely a result of his familiarity with the "Zodiac Man," a didactic medical-astrological tool that superimposed emblems of the zodiac on anatomical drawings to show their corresponding points of influence, as can be seen in figure 4.[3] An analysis of God's work through secondary causes such as celestial influence—and of the distinctions between natural events and miracles—was of particular interest to scholastic theologians in the central and late Middle Ages. Anselm of Canterbury distinguished seemingly miraculous natural events from actual miracles by dividing phe-

Figure 4. The Zodiac Man, a medical guide to the influence of zodiacal signs over parts of the body. *Source:* Oxford, Bodleian Library, MS. Savile 39, fol. 7r.

nomena into three categories: the natural, the voluntary (done in accordance with human will), and the marvelous (here designating what would later be termed miraculous). The ontological and terminological differences between "*mirabilia*" (natural wonders or marvels) and "*miracula*" (miracles) were further elaborated by university intellectuals in the thirteenth century, along with the presumption that although nature operated for the most part autonomously, it was in alignment with God's will.[4]

Rupescissa's likening of God and nature to craftsman and tool does not exhaust his repertoire of images with which to describe nature. Nature is not only the artisanal tool within God's workshop; it is also a book upon which he writes a providential script. Nature is therefore analogous to the Book of Scripture and, like it, a means to uncover apocalyptic prophecy.

{131}

Strikingly, Rupescissa argues that the prophetic information embedded in scripture can also be gleaned from observations of the natural world. One has only to study the movements of the stars and planets at significant times in order to read current and future developments in sacred history. The Book of Nature and the Book of Scripture are thus parallel sources: both are matter imbued with divinity and, in both cases, an understanding of their eschatological content is contingent upon God's will to reveal himself to readers. Furthermore, Rupescissa identified *concordiae*, or agreements, between nature and Scripture that worked in much the same way as the *concordiae* identified by Joachim of Fiore in his eschatological reading of the Bible. Nature, like scripture, was a repository of clues left by God about the events of the last days. With the right exegetical method, the reader could decipher them and acquire advance knowledge of the end.

In the *Liber ostensor*, Rupescissa performed a complicated exegesis of an otherwise unknown prophetic text with the incipit *Cum necatur flos ursi*.[5] Using the information contained within the prophecy, Rupescissa attempted to calculate the exact dates of apocalyptic events, including the commencement of apocalyptic "tribulations" and the final appearance of Antichrist. He confirmed his conclusions by comparing *Cum necatur flos ursi* (and other prophecies) to two other sources of prophetic information: divine scripture and observations of natural phenomena, in particular, the positions of stars and planets at critical times. Rupescissa argued that this method of comparison allowed him to determine accurately the specific dates of the catastrophes to come.

The book begins with a word-for-word exposition of the *Cum necatur* prophecy, which, in a claim to mystical authority, Rupescissa said was first revealed to him by the Virgin Mary in a vision.[6] Much of the section consists of quotations from the prophecy followed by commentary situating the predictions within a contemporary political context. Rupescissa identified a number of *Cum necatur* prophecies as references to current political leaders and to recent military conflicts. For instance, he interpreted the phrase "death of the bear" as symbolizing the death of Andrew of Hungary in 1345; similarly, he identified the prophecy's "new Alexander" as Louis of Tarento (d. 1362), Andrew's successor as king of Naples and Sicily.[7] After his demonstration of the correspondences between *Cum necatur* and current events, Rupescissa used the text to predict the future political affairs and natural disasters that he believed would accompany the reign of Antichrist. For example, a passage from the *Cum necatur* prophecy states that "tempests, floods, barrenness, intemperance of season, [and] corruption of the air" will be seen.[8] Rupescissa interpreted this passage as a reference to the natural disasters and deprivations that Christendom would soon face as the end of time neared.

He bolstered the validity of his predictions by comparing the *Cum necatur* prophecy to evidence from astronomical observation.[9] He dated the beginning of the tribulations to the period between 1360 and 1365, basing his calculation upon the great conjunction of the three superior planets (Saturn, Jupiter, and Mars) that preceded the appearance of the Black Death in 1345 and that was scheduled to terminate its influence in 1365. During 1365, a great conjunction of the two superior planets (Saturn and Jupiter) would pass from the airy "triplicity" to the watery one, entering the aquatic sign of Scorpio.[10] The two superior planets would remain in the same triplicity for a total of twelve conjunctions before they moved to a new triplicity. Rupescissa reported that the last time this had happened was in 1125. He therefore concluded that beginning in the years leading up to the conjunction of 1365, a variety of natural disasters would take place on earth: plagues would spread, corruptions of air would disturb the atmosphere, and, finally, Antichrist himself would appear. Rupescissa seems to suggest that, much like an exegetical reading of revealed Scripture, observation of the natural world was a means to secure apocalyptic prophecy.

Rupescissa's attempts to link planetary conjunctions to religious history, and particularly to the rise of Antichrist, were closely related to the astrological theories of the Arab Abū Ma'shar (known in Latin as Albumasar, 787–886) in his *On Great Conjunctions*, an extremely popular astrological text. The treatise was the major source of planetary-conjunction theory in the Middle Ages, and it delineated the pattern of great conjunctions as they moved through the signs of the zodiac, a system that was adopted by Rupescissa. Abū Ma'shar's theory of astrological history taught that six principal changes in religious or governmental dynasties had happened or would happen in the future. These changes corresponded to conjunctions of Jupiter and Saturn occurring at intervals of 240 or 960 years, that is, at the time of a "greater conjunction" (the shift of a series of planetary conjunctions to a new "triplicity" of zodiac signs) or at the "greatest conjunction" (when the conjunctions once more returned to the sign of Aries).

Zodiacal shifts of triplicity were an essential concept within conjunction theory. In brief, Saturn completes a course through the zodiac in approximately thirty years, Jupiter in approximately twelve years. Thus, a planetary conjunction between Jupiter and Saturn occurs roughly every twenty years, each time in a different position within the zodiac. For example, if the first conjunction of the two planets occurs in the sign of Aries, the second will be in Sagittarius, the third in Leo, and the fourth once more in Aries. But because the planets do not complete their courses in precisely thirty and twelve years, after 240 years, the next round of conjunctions will shift slightly from the point of the planets' initial conjunction. Subsequent conjunctions will fall within the same three zodiacal signs (the triplicity), but they will slowly progress until a conjunction falls in a new sign, beginning a new triplicity.[11]

Conjunctions falling in a new triplicity were thought to have dramatic consequences for life on earth. Abū Ma'shar speculated that such conjunctions could be used to predict future revolutions in world domination, as well as events of lesser importance. For example, according to his theory, a shift in triplicity in 571 heralded the rise of Islam and the political dominance of the Arabs; a similar shift in 749 was linked to the Abbasid revolution. The political astrology of Abū Ma'shar played an important role in millennial movements within the Islamic world, since reports of an approaching shift in triplicity could stimulate expectations of radical change in religious and governmental institutions. News of the shift of 908 and 909 prompted the millennial Ismāʿīlī missionary ʿAbū ʿAbd Allāh al-Shīʿī to establish the Fatimid state in North Africa in 909. An erroneous prediction of a conjunction in 929 led to apocalyptic fears in Bahrain that the era of Islam and the political domination of the Arabs would soon come to an end.[12]

Abū Ma'shar and his followers linked the six changes in religious and governmental regimes to six celestial bodies in conjunction with Jupiter: Judaism was associated with a conjunction of Saturn; pagan religion, with Mars; worship of stars and idols, with the sun; Islam, with Venus; Christianity, with Mercury; and a "lying sect," with the moon.[13] Rupescissa's application of conjunction theory to Christian apocalypticism was not new. Roger Bacon had already adapted Abū Ma'shar's system to the context of thirteenth-century Europe and offered his own version of conjunction theory in the *Opus maius* and *Opus tertium*. According to Bacon, each of the six religious sects was again associated with a planet in conjunction with Jupiter: the religion of the Jews was linked to Saturn; the Chaldeans, to Mars; the Egyptians, to the sun; the Saracens, to Venus; and the Christians, to Mercury. The final conjunction between Jupiter and the moon represented the sect of Antichrist.[14]

Bacon was so convinced of the utility of interpreting prophecies and astrological phenomena for predicting the apocalypse that he urged Pope Clement IV to adopt such a program of study for the church. In the *Opus maius*, Bacon claimed:

> I know that if the church should be willing to consider the sacred text and prophecies, and also the prophecies of the Sibyl and of Merlin, Aquila, Seston, Joachim, and many others, and above all the histories and books of the philosophers, and should order a study of the paths of astronomy, it would gain some greater certainty regarding the time of Antichrist.[15]

As we have seen, astrology was just one part of Bacon's scientific approach to the apocalypse. Through a combination of alchemy, astrology, and species multiplication, Bacon planned to alter the atmosphere of the earth and make conditions favorable to defeat Antichrist, convert the infidels, and reform Christian society. For Bacon, astrology was instrumental for predicting and preparing for future events associated with the coming end.

In contrast, many Christians found astrological concepts difficult to reconcile with their religion. The determination of terrestrial events by celestial bodies conflicted with the Christian emphasis on human free will and divine omnipotence. Moreover, the astrological premise of the Great Year (the time after which all stars would return to their original positions, causing all previous events on earth to repeat) was clearly at odds with the Christian narrative of progressive history.[16] Christian philosophers attempted to harmonize astrology and free will in various ways. Augustine's *De civitate Dei* claimed that "stars give notice of events and do not cause those events, so that their position becomes a kind of statement that predicts, but does not produce, the future."[17] By the thirteenth century, the standard view had developed to allow astrology power over the body but not the soul. It was thus difficult, but not impossible, for human beings to resist the movements, rays, and "influence" (*influentia*—an invisible flow that also affected magnetism, tides, and other natural phenomena) of celestial bodies.[18] Nevertheless, theologians such as Thomas Aquinas and Bonaventure, who assumed a causal role for celestial influence, were more comfortable with "universal" astrology, which determined the fate of large regions or communities rather than individuals.[19]

Astrology was frequently used to predict such large-scale events, including, as we have seen in the writings of Roger Bacon, the end of time. But after Bishop Tempier's commission of 1277 included astrological theories among its condemned propositions, skepticism about the efficacy of astrology's predictive power increased.[20] Around the turn of the fourteenth century, a number of theologians and apocalypticists flatly denied the possibility of predicting the eschaton through astronomical observation. Arnald of Vilanova offered a calculation of Antichrist's arrival based on a creative scriptural reading; he claimed that his prediction was superior to those based on astrology because just as God "acted supernaturally in the work of the world's creation, so too will he accomplish the world's consummation supernaturally."[21] Such an explanation appears to deny the possibility of determining apocalyptic times through the study of natural phenomena. But theologians, including John of Paris and Henry of Harclay, found Arnald's prediction to be hardly more palatable than those of astrologers.[22] Harclay dismissed as ridiculous assertions that astrology or any other method of calculation (including prophetic and scriptural analysis) could predict the time of Antichrist's appearance. Human beings were simply not able to have precise knowledge of the time of the end. Laura Smoller has noted that such criticisms had a strong effect on the next generation of astrologers: although astrologers would claim credit for predicting the plague in the mid-fourteenth century, they would avoid attempts to predict the apocalypse.[23]

John of Rupescissa, however, did not shy away from such ambitious calculations. This was in part because he had a more complicated notion of the natural aspects of the end of time. He believed that there would be truly supernatural aspects of the apocalypse, but he reserved a significant role for human involvement in its tribulations through the workings of "perfected nature." This may have prompted him to make a bold astrological prediction of the end. Rupescissa demonstrated familiarity with Abū Ma'shar's (and possibly Roger Bacon's) connections between astronomy and religious sects, particularly as they related to predictions of Antichrist's arrival. He thus wrote about the great conjunction of 1345:

the philosophers say that this conjunction signified that within twenty years in the future, new future sects and horrible disputes between the faiths would most certainly occur. According to the rules given by the ancient philosophers and in writings left behind on the year 1345, all Saracen and Jewish and Christian astronomers writing about the influence of the aforementioned conjunction unanimously, without dissent, judged and put down in writing that because Saturn by nature signifies the Jews and the law of the Jews (as Albumasar wrote in the first treatise of the book on the conjunctions and revolutions of the years of the world), and because [Saturn] was stronger than Jupiter and Mars in its house and triplicity, this designated that within the twenty years that end in the year 1345, an endeavor on the part of the Jews and their false messiah would have to happen.[24]

Rupescissa thus linked the great conjunction of 1345, in which he claimed Saturn dominated, to the enrapture of the Jews by the Jewish false messiah, whom Rupescissa identifies elsewhere as the Antichrist of the East.[25] In the above passage, Rupescissa makes use of Abū Ma'shar's system of historical astrology, as well as the adaptation of Abū Ma'shar to medieval Christian eschatology.

Through his analysis of planetary conjunctions and their effects, Rupescissa concluded that readers should soon expect the appearance of Antichrist and a number of natural disasters. His astrological predictions seemed to him to corroborate the *Cum necatur* prophecy, which warned of "tempests, floods, barrenness, intemperance of season, [and] corruption of the air." Such agreement in the two sources demonstrated for Rupescissa the accuracy of his own interpretations.

Finally, he argued that his predictions based upon astronomical evidence and the *Cum necatur* prophecy could be confirmed by an eschatological reading of the Book of Job. Rupescissa wrote that in Job 41 the same natural disasters were symbolized by the figure of Leviathan, which preceded the appearance of Antichrist:

[A]bout Antichrist God thus says, 'his breath kindles coals and a flame goes out from his mouth; in his neck strength will dwell, and want precedes his face.' (Job 41:12–13) If, therefore, want precedes his face, then want arises through tempests and floods and intemperance of season and corruptions of air, and deaths arise through wants. This clearly follows the marvelous concordances between the nature of the influence of the stars and the heavens and this [*Cum necatur*] oracle and divine scripture.[26]

The agreement of astronomical observation and textual prophecy on the advent of Antichrist demonstrated for Rupescissa a "concordance (*concordiam*) between nature and divine scripture for the aforementioned five pestiferous years."[27] Rupescissa's use of the word "*concordia*" to describe this agreement is important: for an avid reader of Joachim of Fiore (and the works attributed to him) the term had special meaning. In the Joachite apocalyptic schema, *concordiae*, or concordances, were the eschatological correspondences between the Old and New Testaments, which referred to each other and ahead to the events described in John's Apocalypse.[28] Bernard McGinn has identified *concordia* as one of the two keystones of Joachim's exegetical method, demonstrating the centrality of the concept to a Joachite reading of sacred texts and sacred history.[29] Clues to the end of time mercifully left for human beings by God were thus not only found in the traditional forms of revelation (that is, Scripture and prophecy) but also in the natural world that surrounded humanity. The study of nature could reveal information about the precise time and character of the last days, giving humans a means to prepare for Antichrist. The significance of *concordia* to Rupescissa's reading of nature is also related to his experience with the scholastic method at the University of Toulouse. *Concordantia*, or the reconciliation of contradictory authorities (particularly of Scripture and philosophy), was a central aspect of scholasticism, and Rupescissa's discussion of the *concordia* between nature and textual prophecy reflects a similar desire for agreement and clarification. Rather than reconciling the contradictions between philosophical authorities and divine Scripture, Rupescissa attempted to show how natural philosophy agreed with the eschatological information contained in the Bible and in prophetic treatises. His emphasis on the role of *concordiae* reveals the influence of both the Joachite apocalyptic schema and the common method of analysis used by scholastics during this period.

Rupescissa's use of the word "*concordiae*" to describe agreements in nature and divine Scripture furthermore suggests that nature could be read in the same fashion as a book. It was outfitted with the same sort of textual correspondences and eschatological meanings as divine writings (including the Bible and apocryphal prophecies, both of which Rupescissa considered to be divinely revealed), and it contained a similar cache of hidden prophecy. Nature was thus a book as divine and rich in meaning as holy Scripture. This metaphor of the Book of Nature is perhaps not surprising, considering its frequent use in literary and theological explorations of nature, especially those associated with the heightened interest in the natural world that emerged in the twelfth and thirteenth centuries.

Augustine, Hugh of St. Victor, and Bonaventure all conceived of nature as a book in their writings.[30]

Rupescissa's *concordiae* between the Book of Nature and the Book of Scripture also underscored the agreement of the two texts on the tribulations of Antichrist. Predictions of the end based on the Book of Nature and the Book of Scripture were possible because of the "marvelous concordances" among nature, Scripture, and prophecy, which alluded to one another and explained the future events of the apocalypse. Unity of texts was a significant element of Joachim of Fiore's exegetical schema, which drew upon the cohesion of the Old and New Testaments in their capacity to point to each other's contents and to illuminate the meaning of the Book of Revelation. This process is strikingly similar to Rupescissa's, although Rupescissa added the Book of Nature to the fundamental texts upon which an eschatological reading of history could be based. Rupescissa's exegesis of nature and Scripture deemphasized the distinctions between the word of God and the world of his creations, and it argued for a richer sense of the divine in all natural processes. Moreover, concordances rationalized the study of astrology (and by extension, the natural world) since the fruits of such learning could be confirmed by an exegesis of divine Scripture. Such accord served to place the natural world within a teleology of history.

For Rupescissa, then, the task of the alchemist-exegete was to learn the causes of natural wonders—the workings of alchemical medicine, the coordinates of the planets and the stars—to recognize and manipulate the elements of perfected nature within them. Nature expressed the ineffable will of God, who wielded it as a craftsman wielded a tool. The mechanical movements of the celestial bodies revealed sacred prophecy about the advent of Antichrist and his apocalyptic tribulations. In apocalyptic exegesis, as in alchemy, the perfected nature of the celestial sphere—and the movements of its bodies, the stars and planets—allowed human beings to grapple with the final events of the age. Nature provided clues to such events and suggested the countermeasures that people could take to avoid them.

Scholars such as Laura Smoller and R. W. Southern have written about efforts in the thirteenth century to "naturalize" the apocalypse by offering natural philosophical explanations of supposedly supernatural apocalyptic phenomena.[31] Smoller viewed this as part of the larger trend among medieval scholastics to explain marvels in terms of natural causes. This program extended logically to the events of the last days as described in prophecies and Scripture. She argues that this naturalizing impulse served to draw the events of the apocalypse—by definition incomprehensible to humanity—into the realm of human understanding. At the turn of the fourteenth century, however, authors such as Arnald of Vilanova, Henry of Harclay, and John of Paris suggested that apocalyptic events would occur supernaturally, rather than naturally; they therefore deemed naturalist approaches to the apocalypse to be unhelpful. But by the mid-fourteenth century, explanations of marvels such as plagues and earthquakes were often ambiguous, including language that was both naturalist and apocalyptic. According to Smoller, this ambivalent approach revealed the hesitancy of the authors to announce the end, a trend that she believes led to renewed speculation about the natural causes of seemingly apocalyptic phenomena.[32]

Rupescissa differs somewhat from the natural philosopher-*cum*-apocalypticists that Smoller describes in her article. While writers such as Heinrich of Herford and John Clynn expressed caution by describing marvels as simultaneously natural and apocalyptic, Rupescissa was clearly convinced of the imminence of the apocalypse and its direct relationship to virtually all the social, political, and natural phenomena that he discussed. His was not an effort to prudently "predict and not predict the end at the same time," as was the case for some of his contemporaries.[33] Instead, he grasped at natural philosophy as one of many tools with which to analyze the crisis. His conclusions were a source of consternation for at least one critic, the French astrologer Simon de Phares (fl. c. 1490), who praised Rupescissa as "a most subtle spirit and a great speculator in the science of the stars" who nevertheless strayed beyond the limits of astrology, confusing fantasy with divine revelation.[34] De Phares's denigration of Rupescissa's astrology echoes the judgment of Rupescissa's scriptural exegesis by the papal curia more than a century earlier in 1349.

For Rupescissa, natural philosophy not only clarified the apocalyptic; the apocalyptic clarified nature. Rupescissa's interest was not in describing and explaining events associated with the last days, such as monstrous births, plagues, or rains of toads. Instead, he showed how the apocalyptic meaning of time and history was expressed in and acted upon by the processes of the natural world. He also presented natural marvels, alchemy, and astrology as ways for humans to cope with the apocalypse. Rupescissa's efforts thus might be considered an attempt to "apocalypticize" the natural world. In his works, apocalypticism provides the metaphorical language to describe nature's biological and chemical processes. Moreover, it situates the natural world within an apocalyptic framework and gives human beings an active role in history. Rupescissa's writings therefore do not demote the supernatural to the level of the natural and comprehensible. Nor do they express doubt about the immediacy of the apocalypse. Instead, they elevate nature to the significance of theology. Nature has a role in the course of salvation history and its culmination in the apocalypse. Its study is no less crucial.

The Natural and the Artificial

Although Rupescissa praised nature and pondered the transcendence of natural wonders, he also expressed a lack of faith in nature's ability to manufacture its products flawlessly. By pointing out (corrupted) nature's inadequacy, Rupescissa provided a rationale for the art of the alchemist as the helper or corrector of nature. A close reading of Rupescissa's alchemical writings reveals a fascinating interpretation of the relationship between nature and human craft. On the one hand, Rupescissa admitted the inability of human craft to rival the products of the natural world. On the other hand, he heralded the alchemist as a worker of "miracles" to the "cure" of nature.

In one passage of *De quinta essentia*, Rupescissa portrayed the alchemist as a mere imitator of nature who, despite his efforts, could not match the quality of natural creations. Rupescissa warned that although the human-made quintessence was pure and incorruptible, and although its actions mirrored those of the fifth element in the heavens, quintessence "is not entirely brought to the incorruptibility of heaven, just as artifice does not approach nature. But nevertheless it is incorruptible in comparison to a composition made from the four elements."[35] Such a distinction between the quality of the manufactured quintessence and that of the natural heavenly quintessence was necessary, Rupescissa argued, because if quintessence were totally incorruptible like the heavens, it would preserve the human body forever, against the wishes of the "creator of human nature," Jesus Christ. Human artifice was therefore limited in its ability to produce the incorruptible quality of the heavens, just as artifice was unable to equal nature.

Although Rupescissa maintained that artificial things were inferior to natural ones, he also suggested in several passages of *De quinta essentia* that alchemists could serve as conduits for a more perfect form of nature and, as a result, improve upon corrupted nature. In Rupescissa's discussion, nature is the model for the art of the alchemist, but, in some cases, artificial products may actually be superior to natural ones. Rupescissa's instructions for preparing the "quintessence of blood" reveal this paradox. According to Rupescissa, the quintessence had to be "extracted through human artifice from the body of nature created by God."[36] Although the quintessence was made by God in the natural world, it still needed to be obtained and refined by means of human art. Both stages of preparation—natural and artificial—were necessary to complete the quintessence and receive its benefits. Rupescissa explained that the quintessence of blood could be used to heal wounds and cure illness because "this extraordinary quintessence is the greatest thing of nature to be had, since in it is the marvelous virtue of our starry heaven and it performs the most divine miracles to the *cure of nature*, just as I will teach below."[37]

Rupescissa here alludes to the "perfected nature" of the quintessence, which has the "marvelous virtue of our starry heaven." Perfected nature perfects the quintessence so that it is able to bring about marvelous cures; this idea asserts the power of a higher order of nature that operates in the realm of alchemical medicine. We may see here Rupescissa's two levels of nature: corrupted terrestrial nature and perfected celestial nature. He suggests that the quintessence can lift the former to a higher level of perfection by repairing the damage done to the body by age and disease. Through alchemy, human artifice may thus work a "divine miracle" that actually *cures* corrupted nature.

Rupescissa's assertion of both the impotence and efficacy of human craft points to a larger debate in alchemical literature between those who maintained the superiority of natural products to artificial ones and those who made novel, even boastful claims about the alchemist's power to outperform nature.[38] The debate stems from the famous remarks of Avicenna (al-Husain b. ʿAbdallāh Ibn Sīnā, 980–1037) in *De congelatione et conglutinatione lapidum* (the passage is commonly known by its incipit, *Sciant artifices*), a work on the formation of metals that acquired both great authority and a widespread audience when it was appended without attribution to an early-thirteenth-century translation of Aristotle's *Meteorologica*.[39] Avicenna's text argued that the alchemical transmutation of metals was not possible because artificial things were inherently inferior to natural ones. Although a base metal may seem to be transmuted because of accidental or superficial changes in its appearance, its original species remained intact. As a result, an alchemist could not hope to change the species of a metal. Even if a metal appeared to become gold, Avicenna warned that such change was impossible because "art is weaker than nature and does not equal it, however much it labor[s]."[40]

Avicenna's position differed from that of the true Aristotle, who allowed in his *Physica* that human art might mimic or even improve upon nature. First, Aristotle distinguished between natural and artificial products by noting that the natural has "within itself the principle of its own making," while the artificial arises through "some external agent."[41] Second, Aristotle identified two kinds of art: one that "imitates" nature and another that "perfects that which nature cannot complete."[42] The Aristotelian view clearly offered to alchemy greater latitude for artifice to rival the processes of nature. Despite its divergence from Aristotle, *Sciant artifices* enjoyed the Philosopher's imprimatur until its authority was disputed by Roger Bacon and Thomas Aquinas in the late thirteenth century, and it remained popular even after its true author was established.[43] William R. Newman has suggested that the great influence of *Sciant artifices* led its opponents to adopt equally radical views of humanity's ability to equal, or even surpass, nature.[44] Roger Bacon, for instance, claimed that things are created "better and more plentifully by art than they are produced by nature," while Paul of Tarento, the actual author of the popular *Summa perfectionis*, speculated that human technology was capable of manufacturing anything short of the animated body infused with a soul.[45]

In his analysis, Rupescissa was perhaps influenced by the Aristotelian interpretation of human art as capable of "perfect[ing] that which nature cannot complete," a sentiment also expressed by natural philosophers such as Albert the Great and Petrus Bonus of Ferrara. This allowed a role for artifice not in the creation of things but in the aid or the correction of nature's creation. Albert explained the process:

Nature itself performs the work, and not art, except as the instrument, aiding and hastening the process, as we have said. . . . And whatever nature produces by the heat of the sun and the stars, art also produces by the heat of the fire, provided the fire is tempered so as not to be stronger than the self-moving formative power in the metals; for there is a celestial power mixed with it in the beginning, which may be deflected towards one result or another by the help of art.[46]

According to this view, human art could facilitate or accelerate the natural formation of metals, but nature was actually responsible for accomplishing the task.[47] Albert speculated that the natural celestial heat that formed metals could be imitated by a human-made fire used in alchemical operations. The artificial heat would assist in and enhance their natural development.[48] Petrus Bonus made a similar point in his *Pretiosa margarita novella*, writing that the artificial methods of alchemy encouraged natural processes and effects.[49] Transmutation was thus artificial, inasmuch as it required human action, and natural, inasmuch as it followed the normal processes that would eventually occur over time.

Comparisons of alchemy to accepted forms of human intervention in nature, such as plant grafting and medical treatment, were common in thirteenth- and fourteenth-century alchemical texts. Albert the Great, for instance, argued that alchemists were to metals as physicians were to patients. Both purged impurities from bodies then strengthened the bodies through medicines designed to encourage natural healing.[50] Following this reasoning, art was able to assist nature because it used nature's own processes. A description of this sort of imitation also appears in the *Meteorologica*, in which Aristotle draws parallels between concoction through natural body heat, such as that which occurs during digestion, and the artificial cooking of food through boiling or broiling over a fire.[51] The influence of this analogy prompted medieval writers such as Albert to theorize that alchemists could transmute base metals by "cooking" them, or applying the artificial heat necessary to attain the equilibrium characteristic of precious metals.

Such analogies used in defense of alchemy helped to establish a place for human intervention in metal formation. They also showed a gradation in natural objects, which could become more or less corrupt through human manipulation. Rupescissa asserted that the red and white philosophers' stones converted base metals into "perfect silver and gold, better than that which comes from the mines."[52] Rupescissa thus claimed that human intervention could lift the metals of corrupt nature to the higher level of perfected nature. Other alchemists had already made similar claims: Petrus Bonus wrote that gold created through alchemy was superior to natural gold because alchemy purged the metal of all its sulfurous impurities, which still remained in small quantities in natural gold. Because of this, natural gold was theoretically capable of reaching a higher degree of perfection. Alchemical gold, in contrast, was at the highest level of perfection possible for a metal. Petrus added that human artifice might give to gold a more intense and attractive coloring than was naturally possible.[53] Related theories about the superiority of art to nature appear in other alchemical treatises of the thirteenth and fourteenth centuries, as we have seen from the spirited claims of Roger Bacon and Paul of Tarento. Their assertions placed the entire created world under the domain of human art, which they believed manufactured products (with the exception of sentient beings) more quickly and more effectively than nature was able to achieve.

This climate of optimism about human technology fostered new claims for the potential benefits of alchemy, such as Roger Bacon's belief that elixir could restore the perfect complexionary balance enjoyed by both the biblical patriarchs and resurrected bodies at the end of time.[54] In both Bacon and Rupescissa's writings, alchemy functions as a basis for many different fields of knowledge and technology, and even as a weapon of sorts in the future battle against Antichrist. Bacon was so enthusiastic in his praise of alchemy that he labeled it the root of all natural philosophy, the field of learning he believed to be humanity's best defense against Antichrist during the last days.[55] Rupescissa adopted Bacon's optimistic view of alchemy and inserted into it the innovative category of perfected nature; this made it possible to clarify how artifice could outperform nature. He argued that the alcohol-based quintessence was actually a piece of perfected nature that could return the human body to its formerly youthful state and preserve it indefinitely from age and disease by "curing" its corrupted nature.

There was also, of course, a counterargument to this optimism. One version of the Pseudo-Aristotelian *Secretum secretorum* warned that it is "impossible to know how to produce genuine silver and gold, since it is impossible to become the equal of God the Highest in his own works."[56] This passage reveals a strong reaction to the arrogance of alchemists eager to trespass upon the creative powers of divinity. Certain scholastic naturalists, including Thomas Aquinas and Giles of Rome, continued to adhere to the original tenets of *Sciant artifices* and deny either that transmutation was possible or that transmuted metals were equivalent to natural ones. Both Aquinas and Giles argued that alchemical gold was deficient in comparison to natural gold, since the conditions in the bowels of the earth necessary for proper metal maturation could not be replicated artificially.[57] Alchemical gold therefore did not possess all of the qualities of natural gold and could actually be harmful to those who ingested it as medicine. Rupescissa was no doubt persuaded by this point of view: he compared alchemical gold unfavorably to other forms of gold in *De quinta essentia*, writing that the "gold of man, which is of alchemy" is noxious and rife with corrosives so that it "destroys nature."[58] He did not, however, consider gold made by his own recipe to be alchemical gold. Instead, he called it the "gold of God," and deemed it the best gold for use in quintessence.[59] He was moreover careful to preserve the ultimate power of God over human longevity by emphasizing that alchemy could not extend life beyond the point of divinely preordained death. In this discussion, we may see traces of reactionary skepticism about alchemy interwoven with Baconian optimism about the power of human artifice.

But how could human art improve upon nature? And why was nature subject to human power? Paul of Tarento and Petrus Bonus argued that human mastery over nature resulted from nature's subordination to the human intellect.[60] The idea that the natural world was subject to human manipulation—subordinated to humans by God in the Book of Genesis account of creation—was often repeated throughout the Middle Ages, particularly in twelfth-century reflections (such as those of Gilbert de la Porrée and Alan of Lille) on human craftsmanship and the place of human-made products among God's creations.[61]

Rupescissa agreed that nature was subject to human will, but he offered a somewhat different analysis of artifice's superiority to nature. Rupescissa described a two-part process in the creation of quintessence. The first part was completed by perfected nature and the second by human artifice, which, according to Rupescissa, intervened in natural processes on behalf of God. He explained that humans were able to arrest the deteriorations of the body from age and disease through the quintessence because of "the virtue which God contributed to established nature and subjected to human magisterium."[62] Rupescissa upheld the standard view that God produced the natural world and submitted it to human rule. He added, however, that the action of the quintessence was actually a "divine miracle" performed through the conduit of human art: "with our Sun and the terrestrial stars you [the alchemist] make the work of God correctly and, by his power, miracles over the earth."[63] Alchemy presented humans with an opportunity to channel the power of perfected nature in order to "cure" or "correct" the deficiencies of corrupted nature. This provided a rationale for artifice's superiority, since perfected nature could certainly improve upon things engineered by corrupted nature alone. Rupescissa emphasized this celestial power of human art in his *De quinta essentia*, writing that his theoretical exposition of alchemy in book 1 revealed to readers "the heavenly principles of artifice."[64] Such language encapsulated the melding of these two aspects of alchemy: the manipulation of nature through human art and the celestial virtue from which this art received its force.

The alchemist's transmutation of metals was therefore a microcosm not just of Christ's passion, as I demonstrated in chapter 6, but also of creation, since new forms were generated by means of perfected nature through the conduit of the alchemist. The parallels between the alchemist's operation and God's creation of the natural world dramatize alchemy through its relationship to the central events of salvation history—the Beginning, the Incarnation, and the End. The alchemical knowledge needed to channel perfected nature would take place in the context of the last days, which Rupescissa thought were already happening. During this time, according to Joachite predictions, humanity would enter the third state of history and become illuminated with the Holy Spirit. Alchemists were therefore learning the means to channel the celestial world onto earth at the same time that pious Christians were suddenly beginning to comprehend the mysteries of divinity.

Although the cures achieved by the quintessence were natural, for Rupescissa, nature was in need of assistance—a "handmaiden," as Paul of Tarento suggested—and alchemy, which allowed humans to reach perfected nature, could provide it. According to Rupescissa, perfected nature was the source of the quintessence's incorruptibility, and it made possible the repair of corrupted nature and the defects that resulted from its operations. This scenario demonstrates Rupescissa's tendency to bring together the usually separate spheres of the terrestrial and celestial. Human artifice worked in concert with perfected nature, which it assisted. It also beat corrupted nature at its own game, performing better and faster by means of the celestial power in the alchemist's operation. Artifice thus allowed the alchemist to work "miracles"—acts against the ordinary course of corrupted nature—in God's name and virtue. Rupescissa's model highlights the commonalities of humanity, terrestrial nature, and celestial nature, rather than enforcing a rigid separation between their domains of operation.[65] Each of the three powers works through or on behalf of the others, and each is aimed toward the same objective: the raising of the cosmos to a higher level of perfection. According to this model, the realms of the earth and heaven are blended. There is something of both the natural and the divine in the craft of the alchemist. As the quintessence is prepared and human bodies are fortified for the battle against Antichrist, the various spheres of the cosmos—the natural terrestrial world, the divine celestial world, and the intermediary world of humanity, both material and spiritual—come together.

CONCLUSION

Rupescissa's discussion of nature and artifice reveals how he thought alchemy could address some of the main problems of the apocalypse. Through alchemy, humankind could harness celestial power to work "miracles," producing cures that would prolong life and promote the activities of the evangelical men. This understanding of alchemy allowed Rupescissa to argue that the practical, material activities of the alchemist actually had spiritual import. They promoted a communion of heaven and earth, and their effects figured into the outcome of the apocalyptic crisis. Through this argument, Rupescissa elevated the vocation of the alchemist to that of a miracle worker and protector of Christendom during the last days. Alchemy's significance thus lay in its religious function. It was both a practical art aimed at the production of actual chemical substances and a tool with which to intervene in the spiritual fate of the Christian community. Rupescissa invoked the impending crisis to illustrate humanity's need to overreach the normal operations of corrupted nature and consequently to repair the earth.

This claim is consistent with Rupescissa's faith in the human potential to affect the course of the apocalyptic battle through prophecy and biblical exegesis. According to Rupescissa's arguments in the *Vade mecum* and the *Liber secretorum eventuum*, among other texts, human beings have great power to act in the events of the end and a special obligation to do so. His arguments about human artifice in the realm of alchemy mirror his urgent suggestions that Christians arm themselves with knowledge about the specific dates and events of the end. In *De quinta essentia*, he presses the evangelical men to pursue alchemy in order to repair corrupted nature and lift the terrestrial world to a higher state of perfection. In both Rupescissa's prophetic and alchemical writings—if indeed they can be considered to be two separate genres—he emphasizes the considerable impact of human action. This perspective points to the sheer optimism of Rupescissa's endeavor during what would seem to be one of the darkest periods of European history.

In this chapter, I have explored several different manifestations of Rupescissa's claim that nature operates within the context of a teleological history. Nature has an ontological significance (through its two tiers of perfection and corruption), an operative significance (through its prediction of future events and its capacity to affect them), and an historical significance (through its decisive role in the battle between Christ and Antichrist). As Rupescissa viewed it, nature is imbued with divinity: it reflects the will and the ineffability of God and it contains his plan for salvation. Humans must therefore decipher the meaning and the function of nature in order to survive the future tribulations and bring about the third and final age. Matter, for Rupescissa, was an inherent part of the story of salvation, but it was dependent upon humanity to discover, understand, and manipulate its resources.

Rupescissa imagined a category of "perfected nature" between corrupted terrestrial nature and divine supernature that, through alchemy, could bring earth closer to the level of the heavens. He similarly predicted that earthly society would rise toward the perfection of the theological heaven during the Joachite third state (which also involved a "Repairer" who would purify the corrupted earth). Alchemy and evangelical reform were thus passageways for celestial properties to descend to the realm of the earth and to enter into human bodies. The evangelical men were the bearers of perfection on both fronts. They formed a bridge between this world of generation and corruption and the next world of perfection and quasi immortality. According to Rupescissa, God would act supernaturally in the end, but he also granted to humans the will and agency to bring about through nature immense change in the structure of terrestrial and human society.

Understanding Rupescissa's conception of the role of nature in divine history helps us to observe a direct link between Rupescissa's alchemical and prophetic writings. The Christian apocalyptic provided language with which Rupescissa could discuss complicated and obscure biological and chemical processes. It also situated the natural world within an overarching teleology. Rupescissa portrayed the terrestrial world as bristling with spiritual and eschatological import, and the study of nature drew the inexplicable horror of the apocalypse into the grasp of human comprehension and action. But the world of apocalyptic religion was not demystified by naturalism. Instead, the world of nature was mystified and united with the march of time toward the end.

CONCLUSION

JOHN OF RUPESCISSA'S APOCALYPTIC ALCHEMY

John of Rupescissa survived some of the most tumultuous decades of the Middle Ages. The Great Famine, the Hundred Years War, the Black Death, the Jacquerie, and the papal "Babylonian Captivity" unfolded before his eyes. Viewed through the lens of Joachite and Franciscan Spiritual apocalypticism, such events loomed even larger. To Rupescissa, they constituted mounting evidence that an apocalyptic clash between good and evil was imminent. He saw the persecution of righteous Christians by a church that appeared on the verge of allying with Antichrist. He received news of conflicts among secular and ecclesiastic governments that confirmed the many apocalyptic prophecies he consulted. The onslaughts of war, epidemic, and injustice that he witnessed persuaded Rupescissa that such prophecies, as well as his own visions, were proving all too accurate. In the desperate climate of the fourteenth century, Rupescissa looked to both longstanding and innovative means to understand the crises he observed and to fashion a response to them.

Rupescissa immersed himself in scriptural and extrascriptural prophecies of the end, which provided meaning to the suffering he observed during his lifetime. These apocalyptic prophecies closely mirrored current events, and Rupescissa—like those who lived in earlier and subsequent centuries—was quick to conclude that he was living in the last days of human history. He worked to reconcile preexisting prophecies with the people and events of the fourteenth century, hoping that they might provide clues to future developments. If only humans could determine with a great deal of specificity the *what* and *when* of apocalyptic tragedies, they might shield themselves from the worst of the catastrophes to come. By reacting in this manner to the horrific events of his day, Rupescissa followed a long line of apocalypticists who believed that they were witnessing the climax of history as predicted by the Book of Revelation.

Rupescissa's application of a basically unchanging prophetic structure to contemporary events is, however, only part of the story. I argue that Rupescissa also demonstrates just how nimble and dynamic medieval apocalyptic approaches were. Apocalyptic belief drove recourse not only to timeworn exegetical strategies but also to intellectual and theological innovation. Rupescissa responded to the challenges of the fourteenth century by devising remedies intended to protect humanity from the worst of Antichrist's assaults. Human beings were not merely to watch and wait for future events to overtake them. Instead, Rupescissa highlighted the power of human agency to anticipate the actions of Antichrist, to launch an offensive, and to thwart the enemy. He presented an essentially optimistic apocalyptic program: despite the deaths and disasters soon to arrive, humanity would emerge from the tribulations victorious. Rupescissa seems to have been the first to claim that people could defeat Antichrist and survive into the paradisical third state. The survivors of the coming crisis would populate the postapocalyptic society, renovate the world, and construct a millennium that would last for a literal one thousand years. In the shadows of the mid-fourteenth century, nothing short of a complete purge and restructuring of society seemed sufficient.

Rupescissa's dire predictions of Antichrist and apocalyptic tribulation actually offered something incongruous to his readers: hope. Rupescissa suffered a microcosmic form of the terrible disasters that he predicted for all of humankind in his writings. His conclusions about the causes and effects of the disasters were doubtless shaped by his experiences with the institutional church and his confinement in the papal prison. For someone jailed indefinitely, the present state of the world must have indeed seemed bleak. Yet Rupescissa anticipated a radical change in the near future, and he believed that Christians were meant to play a crucial role in implementing it. The outcome of the apocalyptic battle and the fate of its protagonists were contingent upon the decisions made by people in the heat of the crisis. Because the creation of the new age was the special obligation of Christians, Rupescissa reserved an important place for human action in the present misfortunes. This discourse of activism made the apocalyptic crisis appear more comprehensible and human defeat less inevitable.

CONCLUSION

Like those who participated in the "art-nature" debate of the late Middle Ages, as discussed in the previous chapter, Rupescissa expressed enormous confidence in human beings and their ability to correct and transform the world.

Rupescissa's apocalyptic program was the culmination of a long tradition of Joachite and Spiritual Franciscan prophecy that described an age of the Holy Spirit and a period of divine enlightenment for all people. According to Rupescissa's predictions, the "spiritual" or "evangelical" men who populated Joachite prophecies (and, in his view, who endured the very real attacks of the Franciscan order and institutional church) would be vindicated by the apocalyptic future. Rupescissa's predictions paved the way for those who fought for the ideals of poverty and reform. Moreover, Rupescissa boldly added that because they lived closer to the end of time, the evangelical men's (and his own) interpretations of Scripture were more correct than those of the Church Fathers themselves. In addition, Rupescissa's detailed description of the postapocalyptic millennium emphasized more than ever the rewards that lay on the other side of the battle against Antichrist.

Rupescissa offered to readers not only apocalyptic hermeneutics but also the study of the natural world as a means to combat Antichrist. He identified concordances between the Book of Nature and the Book of Scripture that allowed human beings to read the natural world for eschatological clues in the same manner as a prophetic text. Astrology furnished precious information about the Christian past and future, and it suggested that nature was a credible object of theological study. Alchemy in particular was central to the Christian apocalyptic cause. Alchemy could provide the precious metals and medicines needed to preserve the church and counter the plagues, wars, and other trials that Rupescissa predicted. Why alchemy? Among the naturalist disciplines, only alchemy promised the transmutation of corrupt and imperfect matter. A conventional reading of Aristotelian and Galenic theories left little room for the longevity and earthly perfection that Rupescissa imagined. But Rupescissa was able to draw upon the principles of alchemy to suggest something altogether new: humanity could bypass the normal operations of corrupted nature to manufacture bodies of perfected nature, the nature that more properly belonged to the celestial sphere. At the same time that human beings were called upon to initiate the heavenly third state, they would become capable of channeling the perfection of heaven to earth through distillation.

CONCLUSION

Alchemy and prophecy were thus conduits through which celestial properties could descend to transform both human bodies and the society in which they lived. As Joachite apocalypticism developed and spread during the late Middle Ages, perceptions of the human body and the cosmos were transfigured. The possibility of an apocalyptic purge of society and a future paradisical millennium suggested to Rupescissa that matter, even the corrupt matter of the human body, was ultimately perfectible. This idea had an enormous effect on the way in which he and his audience began to perceive the world around them, and it set in motion Rupescissa's concordance of Christianity and alchemy.

Eschatological fervor not only drove Rupescissa to pursue alchemical, medical, and astrological research, but it also shaped his understanding of those disciplines. In the alchemical processes of transmutation, Rupescissa saw reenactments of the central episodes of Christian history—the genesis, crucifixion, and resurrection—that had the power to intervene in the last days. In addition, he situated alchemy entirely within the context of Christian time: the knowledge to transmute bodies would become available to humanity only as the last days approached. That knowledge, furthermore, derived from the same revelatory insight as Rupescissa's prophetic visions. Rupescissa's writings demonstrate that medieval alchemy should be understood not merely as a forerunner to modern chemistry nor as a product of a Jungian collective unconscious but as a component of fourteenth-century religiosity. By integrating the Christian apocalyptic narrative and the arc of the alchemical opus, Rupescissa engineered a masterful synthesis of Greco-Roman natural philosophy, Arabic alchemy, and Christian spirituality. He viewed what historians tend to interpret as different aspects of the world—such as humanity, nature, and divinity—and different intellectual fields—such as alchemy, medicine, and theology—as permeable and mutually influential. This book is intended, in part, to demonstrate how intellectual and religious traditions interacted to produce new ways of viewing the world and addressing its problems. Rupescissa's body of work reveals a disregard for boundaries between different kinds of scientific and religious thought, boundaries that may, in fact, be an anachronistic creation of later thinkers and modern scholars, who did not or do not view the universe in the synthetic manner that Rupescissa and his fourteenth-century audience did.

Despite the promising new scholarship on alchemy that has appeared in the last two decades, alchemical texts still for the most part constitute unexplored territory for cultural historians. This is unfortunate because alchemy can tell us something important and unexpected about the ways in which medieval people constructed their universe. Although alchemy was never a university discipline, its advocates were among the elites most likely to discuss and determine the boundaries of the natural world. Manuals of alchemy and "books of secrets" that included alchemical concepts were popular among literate audiences, and such texts became even more widely available with the advent of the printing press.[1] Despite a number of prohibitions against the art by ecclesiastical and secular authorities, alchemical research was long supported by royal and papal courts, and many of its known practitioners were physicians, university masters, and even prelates.[2] The alchemical texts of John of Rupescissa, along with those of other theorists and practitioners, serve as valuable windows into the presuppositions of medieval thinkers about social, natural-philosophical, and theological matters. Because alchemy existed alongside sanctioned academic disciplines, borrowing from them and informing them, it provided an alternative view of the natural and social world that had enormous yet generally unacknowledged influence. As we can see in the case of Rupescissa, alchemy offered principles that, although based upon mainstream natural and theological views, actually interrogated and even overturned those views.

While Rupescissa's story is exceptional, he nonetheless reveals a great deal about the culture of the late Middle Ages. His writings point to the manner in which at least some contemporaries reacted to the calamities of the fourteenth century. Rupescissa, his redactors, and his readers responded to the disasters not with mere despair but with vigorous efforts to understand and manage the crisis. A host of novel intellectual and technological tools emerged from this atmosphere. The Christian metaphors and images of Rupescissa's writings, for instance, influenced contemporary views of the natural world and the human body—some of the most foundational components of human experience. It is crucial for historians to examine closely the language and imagery that medieval writers used to express themselves so that we might appreciate the ways in which they built the realities of their worlds. When we can better understand how medieval people imagined the structure of their bodies and the universe around them, we can better see how their imagination derived from both their experiences and their expectations about the societies in which they lived.

CONCLUSION

Rupescissa integrated the natural workings of alchemy and the supernatural workings of Christian history to such an extent that he created a thoroughly Christian alchemy, an endeavor that dissolved the line between theology and naturalism. He used religious imagery to dramatize the workings of the natural world, and he described alchemical processes in a manner that highlighted the presence of the divine within them. Alchemy was not only a mimicry of the passion or of heaven, it was a wonder in its own right, a tool used by God to carry out his will, and a glorification of its creator. Rupescissa rationalized the study of alchemy by picturing the natural world as another way of telling the story of Christianity. Alchemy was useful for providing medical remedies and noble metals, but it was also an expression of providence. By making this connection between nature and divinity, Rupescissa pointed to the study of the natural world as both a spiritual activity and a practical necessity. For Rupescissa, alchemy and theology were by no means distinct activities. Amid the apocalyptic currents of the fourteenth century, eschatological history was imprinted upon nature, and the visible world teemed with invisible meaning. The study of alchemy was thus no different than the study of theology. Both revealed the wonders of God and his plan for humanity and, as such, they pointed to the future humans might create, if only they chose to do so.

THE AFTERLIFE OF THE QUINTESSENCE

Like the apocalyptic heroes of *De quinta essentia*, John of Rupescissa's ideas enjoyed extreme longevity, outlasting their author by several centuries. Although Rupescissa's life has been little studied by modern scholars, his work elicited great interest from his contemporaries and their descendents. The fiery conviction that prompted Rupescissa to speak his mind before his superiors—despite the risk of imprisonment or even execution—convinced at least some that he was indeed divinely inspired. As a result, his travails in prison received a great deal of attention, and they prompted speculation about the origins of his prophecy and the fairness of his prosecution. Chroniclers such as Jean de Venette and Jean Froissart, among others, weighed in upon his fate, and powerful figures petitioned him for his predictions of future events. In the aftermath of his career, Rupescissa's prophetic and alchemical works were copied and disseminated widely,

CONCLUSION

and evidence suggests that a lively discussion of his thought continued well into the seventeenth century.

Manuscript copies of Rupescissa's most popular prophetic work, the *Vade mecum in tribulatione*, proliferated throughout Europe, eventually appearing in many vernacular translations. The text is preserved in more than forty extant manuscript copies, and it was published in a printed edition by Edward Brown in 1690.[3] The *Vade mecum* generated substantial enthusiasm among readers even after Rupescissa's death, particularly during the late fourteenth and early fifteenth centuries. As we have seen, Rupescissa offered a seemingly accurate prediction of the Hundred Years War in 1335, which won him supporters early in his career. Later in his life, Rupescissa predicted that a schism would soon split the church. Following his death, some readers concluded that Rupescissa had correctly predicted the Great Schism of 1378 and was a genuine prophet. Although the critical dates of most of Rupescissa's predictions—which fell mainly in the 1360s—had passed without incident by the late fourteenth century, his followers saw no reason to abandon his prophecy entirely. Instead, they assigned new dates to his apocalyptic predictions and reinvigorated their hopes.[4]

Robert E. Lerner has identified a Rupescissa "reading group" in Saxony in the 1380s led by the Franciscan friar Frederick of Brunswick, who was later to be condemned for heresy along with several of his disciples in 1392.[5] The group, known by their detractors as the "Saxon prophets," borrowed from Rupescissa's prophecies to formulate predictions of a coming "Repairer," a thousand-year millennium, and a heroic role for the Jews in the last days. Dietrich of Arnevelde, who wrote a polemic against the group in 1389, explicitly identified the "nonsense" of "Brother John of Rupescissa" as the source of their delusions.[6] Unfazed by the apparent inaccuracy of Rupescissa's predictions, the Saxon prophets recalculated his dates to arrive at 1399, rather than 1370, as the year of the ultimate defeat of Antichrist. Similarly, they identified the political villain of their time, Duke Otto of Taranto, as Antichrist, whom Rupescissa had earlier identified as Louis of Sicily. The Saxon prophets embraced Rupescissa's prophecies, which they interpreted and modified to fit contemporary events, much as Rupescissa had once interpreted and modified earlier prophecies to fit the events of his time.

The Saxons prophets' contemporary to the south, the Franciscan and prolific Catalan writer Francesc Eiximenis (d. c. 1409), used a similar method to adapt Rupescissa's prophecies to his own ends. Eiximenis,

CONCLUSION

whose influence was greater than his German counterparts because of his close ties to the Aragonese monarchy, drew upon the writings of Rupescissa (whom he referred to as a "great doctor") to craft his own predictions of the thousand-year millennium.[7] Like Rupescissa, his interests encompassed both eschatology and contemporary politics. His theological writings borrowed from the tradition of Spiritual Franciscanism, including the works of Rupescissa and Ubertino of Casale, which he attempted to present in a popular, accessible style. In his encyclopedic volume the *Primer del Crestià* (First Volume of the Christian, 1379), Eiximenis made reference to both the Oracle of Cyril and Rupescissa's commentary on it. Some years later he sent to King John I of Aragon (1387–1395) a letter citing a flattering prophecy of Rupescissa about the house of Aragon.

Rupescissan prophecies circulated widely during this period: an Italian vernacular version of the *Vade mecum* made an appearance in Ciompi-era Florence with its dates amended by an editor to reflect the popular uprising of 1378.[8] A pamphlet written by a pseudonymous author named Telesphorus of Cosenza in the late fourteenth century pointed to Rupescissa's millennial predictions as a pattern for future church reform.[9] Rupescissa's millennial writings were attractive enough in England that they were at one point removed from the libraries of Oxford and Salisbury so they would not be accessible to Lollard dissenters, who admired them.[10] Around 1400, a Bohemian commentator on the *Vade mecum* cited Rupescissa's seemingly correct prediction of the Great Schism and discussed the friar's prophecies on the reduction of the institutional church to poverty. In 1422, another Bohemian cited Rupescissa's prophecies to account for the contemporary political climate of the Hussite rebellion, as well as the burning of Jan Hus in 1415. A Czech translation of a part of the *Vade mecum* also appeared around this time, and it was similarly modified so it appeared to discuss the events of Hussite Bohemia during the 1420s and 1430s.[11] These examples show that Rupescissa's predictions continued to find an engaged audience across a broad geographical area fully sixty years after his death.

While the apocalypticists I have discussed demonstrated little interest in Rupescissa's alchemical writings, at least one writer joined apocalyptic expectation to transmutatory alchemy in a way that suggests he was very likely familiar with Rupescissa's ideas. The anonymous author of the *Buch der heiligen Dreifaltigkeit* (Book of the Holy Trinity), an early fifteenth-century German alchemical text, interspersed predictions of Antichrist with

{157}

instructions for the transmutation of gold and silver.[12] The text demonstrates the author's sympathy for the Franciscan Spirituals' legacy, as well as a strong predilection for the favorite topics of Rupescissa. Like his predecessor, the author of the *Buch* viewed alchemical language as a vehicle for prophecy, and even as a weapon against Antichrist.[13] Although little information about the author is available, we know that he visited the Council of Constance (1414–1418) to promote his work before the Emperor Sigismund and that he attracted the interest of other nobles, including the emperor's brother, Wenceslas of Prague, and the margrave Friedrich of Brandenberg. The *Buch* proved popular enough to be preserved in at least twenty manuscripts, many of which were richly illustrated for wealthy patrons. The survival of Rupescissa's ideas in the *Buch* indicates that a combination of transmutatory alchemy and apocalyptic prophecy continued to find readers at the highest levels of society in the fifteenth century.

Even more influential than his prophecy, Rupescissa's alchemical quintessence became a key concept in the assimilation of alchemical techniques and materials into medical therapeutics during the fifteenth through seventeenth centuries. A large number of manuscripts of Rupescissa's most famous work, *De quinta essentia*, circulated during this period, and over 140 copies are extant. At least fifty-five copies of the *Liber lucis* have also survived.[14] The fifteenth century saw translations of Rupescissa's alchemical writings into English, German, Italian, and Catalan; in the sixteenth century, French, Swedish, and Czech versions appeared. A Latin printed edition of *De quinta essentia* was published by Guglielmo Gratarolo in 1561 and reprinted in 1597, and a French edition by Antoine du Moulin appeared in 1549 and 1581.[15] Gratarolo also edited the *Liber lucis* and inserted it into his *Verae alchemiae* (1561) and *Alchemiae quam vocant* (1572). The *Liber lucis* was published in 1579 under its present name by Daniel van Broekhuizen in Cologne and subsequently reprinted in Basel in 1598 and in Leiden in 1612. Both *De quinta essentia* and *Liber lucis* are included in Lazarus Zetzner's seventeeth-century *Theatrum Chemicum*; the *Liber lucis* also appeared in Jean-Jacques Manget's *Bibliotheca chemica curiosa* of 1702. Rupescissa's theories moreover persisted through the widely read Pseudo-Llullian alchemical works, which culled their ideas from *De quinta essentia* and which were printed at least thirteen times in the sixteenth century.[16] The concept of the quintessence, in both its Rupescissan and Llullian permutations, found audiences eager to possess the marvelous remedies it promised.

CONCLUSION

Lorenzo da Bisticci, an illiterate goldsmith who lived in fifteenth-century Florence, for instance, enjoyed sudden fame as a physician because of his skill with healing waters. Lorenzo's story was well known enough to be recorded in a number of manuscripts, and an account produced in 1462 attributes Lorenzo's knowledge to a text called the *Liber de famulatu philosophie*, another name for Rupescissa's *De quinta essentia*. Bartholomeus Marcellus, who penned the story of Lorenzo's ascent, explained the strange incident:

[Lorenzo] strenuously searched for the great Christ according to the rules of the work *De philosophiae famulatu*—a remedy almost divine and totally unknown today. With God's consent and the help of fortune, he found the Christ of medicine that heals even the helpless sick. Therefore is he revered today as the king of physicians.[17]

The historian Michela Pereira identifies the "Christ of medicine" as John of Rupescissa's quintessence of gold, which was made from gold transmuted by a philosophers' stone. Rupescissa's ideas also found their way into the repertoire of practitioners with more conventional expertise than Lorenzo. The fifteenth-century physician Michele Savonarola, a university-trained master of medicine and a court physician in Ferrara (and the grandfather of the famous Girolamo), drew from alchemical writings on the quintessence. According to his *De aqua ardente*, the substance *aqua vitae* was a "solar quintessence" that had miraculous power, especially over the heart.[18] While Savonarola attributed this idea to the writings of Ramón Llull, it more likely derived from Rupescissa's theories.

Quintessences were in fact to become all the rage in sixteenth-century "iatrochemistry" (or "medical chemistry"), the application of alchemical technologies and mineral-based medicines to human disease. Rupescissa's ideas were rapidly assimilated by a number of authors, including the German surgeon Hieronymous Brunschwygk (1450–1512/13), who cited Rupescissa in his distillation manuals, and the Swiss naturalist Konrad Gesner (1516–1565), who mentioned Rupescissa in his *Thesaurus*. Philipp Ulstad, a professor at the University of Fribourg in Switzerland and the author of the *Coelum philosophorum* (1525), recognized Rupescissa's contribution and substantially developed both the theory of the quintessence and the range of quintessence recipes. Such works had a large readership, piquing the interest of many within the alchemical community, including the iconoclastic Paracelsus (1493–1541), who assured the survival of the quintessence and Rupescissa's legacy through his own popular works.

The Paracelsian alchemical revolution of the sixteenth and early seventeenth centuries, which challenged the Aristotelian-Galenic understanding of the body, was predicated upon the idea of the quintessence, to which Paracelsus devoted part of his *Archidoxis*. Paracelsus's description of the quintessence bears remarkable similarity to that of Rupescissa two centuries before:

The Quintessence therefore, is a certain matter Corporally extracted out of all the things, which Nature hath produced; and also out of every thing that hath a life in its self, and is separated from all impurities and Mortality, is most subtilly mundified, and likewise Separated from all the Elements.[19]

Elsewhere in the work Paracelsus notes that there are quintessences that

Renovate, and *Restore*; that is, such as transmute the Body, Bloud, and Flesh; Othersome for the Conservation of the *Diuturnity*, or Prolongation of the Life; some for the Retaining and Preservation of Youthfulness; some of them work by Transmutation. . . . Briefly, there are many more virtues that they are endued with, which we are able to describe, and their Operations in Medicine, are exceeding admirable and unsearchable, and that variously; for some Quintessences will make a man of 100. years old, like to one of but 20. years of age, and that by their own Vertues and Power.[20]

These claims for the quintessence, which perfected the human body just as transmutatory alchemy perfected metals, resonate with Rupescissa's body of work, as does Paracelsus's belief that the quintessence could rejuvenate the aged, making a man of one hundred "like to one of but 20." Paracelsus was aware of Rupescissa's contribution to the theory of the quintessence, and he mentioned his name with regularity in his works—although he also condemned the treatises of Rupescissa and Arnald of Vilanova in his *On the Correction of Impostures*.[21]

Such interest in the quintessence reached the ears and pen of the French surgeon Ambroise Paré (1517–1590), now best-known for his teratological work *Des monstres et prodiges* (On Monsters and Marvels). According to Paré, by the sixteenth century—fully two hundred years after the composition

CONCLUSION

of Rupescissa's *De quinta essentia*—the idea of the distilled quintessence had become so pervasive that nearly all medicines seemed to be manufactured by this method. Paré describes the fervor for quintessence medicines in his *Oeuvres*:

> [These medicines] are such as consist of a certain fift essence separated from their earthie impurities by Distillation, in which there is a singular, and almost divine efficacie in the cure of diseases. So that of so great an abundance of the medicines there is scarce anie which at this daie Chymists do not distil, or otherwise make them more strong and effectual then they were before [*sic*].[22]

The promise of the quintessence—of healthy and long-lived bodies, of the "divine efficacy" of distillation—clearly exerted a strong influence over practitioners and patients. Aside from the writings of Paré, one can find numerous mentions of the quintessence or its descendants in early modern texts. Recipes collected by the historian Piero Camporesi suggest that medical practitioners preserved quintessences of blood in jars for future use. Camporesi notes that a range of quintessence recipes appear in early modern medical and apothecary texts, many of which enjoyed great success in the sixteenth and seventeenth centuries. The *Della fisica* of Leonardo Fioravanti (1517–1588), for instance, praised the quintessence of wine, as well as a "fifth essence of human blood," with which, Leonardo claimed, "I have as good as raised the dead, giving it as a drink to persons who had all but given up the ghost. Suddenly I have seen them return, and in the briefest time become well."[23] A quintessence of rosemary is among the medicines that survive in the "Giustiniani chest," a Genoese medicine chest of the 1560s now housed in the Wellcome collection.[24]

Paracelsus's followers (known as the Paracelsians) also embraced aspects of Rupescissa's theories and ignited the popularity of quintessences during the period.[25] Not completely unlike Rupescissa's project in the fourteenth century, Paracelsus and the Paracelsians challenged Aristotelian-Galenic approaches to medicine in the sixteenth century, substituting their own salt-sulfur-mercury theory for a regime of treatment based upon the balance of the four humors (Rupescissa, of course, had never lost faith in the basic principles of humoral medicine).[26] Moreover, the Paracelsians asserted the divine purpose at the heart of their studies, eschewing deductive reasoning for an illuminated approach to nature and Scripture based upon experiential research in the laboratory. The Paracelsians were

{161}

not successful in accomplishing a complete reform of academic medicine, as they had hoped, but their efforts led to a greater appreciation of the medicinal value of chemistry, which shaped the later trajectory of pharmacology. Although the Paracelsians ultimately turned their attention from distilled quintessences to residues and precipitates, the medieval technologies engineered by Rupescissa were crucial to their original theories, which survived in some form or other into the seventeenth and even the early eighteenth centuries.[27] While some of the Paracelsians believed that Llull was the originator of their art, Rupescissa's name and genuine works were not forgotten, and curiosity about his writings fueled the production of Rupescissan manuscript copies throughout the period.

Historians of science have recently become interested in the alchemy of the sixteenth and seventeenth centuries, which they now view as a critical period in the development of modern chemistry—an interpretation that has gained traction in the popular imagination as well.[28] Alchemy was the prevailing matter theory during the early modern period, and the experimental approach used by alchemists is now considered one of the hallmarks of modern scientific practice. Certainly, many of the most famous scientists of the early modern period, including Robert Boyle and Isaac Newton, devoted considerable energy to alchemical pursuits. While I have focused on the cultural context of Rupescissa's ideas in the fourteenth century rather than on their correctness or value to modern science, it is worth noting that Rupescissa's quintessence had a dramatic and lasting impact on early modern naturalism. Might we not trace the achievements of sixteenth-century alchemists back to the apocalyptic alchemy of Rupescissa? There is no doubt that the quintessence's legacy can be discerned for centuries after Rupescissa's death, and if modern chemistry owes a debt to the early modern alchemists, then it also owes a debt to Rupescissa. Similarly, we commonly speak of a "quintessence" in modern speech to signify the vital or distilled principle of a substance. This English word derives from the alchemical manuals and poems of the fifteenth century, which ultimately drew their understanding of matter from Rupescissa's writings. Our casual use of the term in the twenty-first century hints at the enduring presence of Rupescissa in our lexicon, just as his alchemical theories are embedded deep within our understanding of chemistry.

Premodern alchemists who borrowed from Rupescissa's naturalist theories often ignored the friar's underlying apocalyptic agenda. But as it entered the mainstream of early modern alchemy, the quintessence may

CONCLUSION

have carried with it more of Rupescissa's innovative and reformist tendencies than historians have imagined. Rupescissa and his quintessence redirected not only the development of early modern alchemy but also the very core of naturalist inquiry. In Rupescissa's works, the horror of change and the desire for changelessness are constant motifs. Yet his writings contained within them the seeds of great historical change. From the kernel of the quintessence arose the Paracelsian revolution, an episode that altered history in a manner far removed from the apocalyptic end anticipated by Rupescissa. But the Paracelsian vision of the future had a distinctly radical Joachite flavor: Paracelsus and his followers advocated no harmonious compromise between old and new; instead, they insisted upon the supercession of previous authorities and bodies of knowledge by the new *scientia* and the new age.[29] Their assault heralded the overthrow of the Aristotelian-Galenic hegemony, the same system of thought that had provided the basis for Rupescissa's theories and to which he had been unfailingly faithful. Perhaps we should see John of Rupescissa as an unwitting agent of intellectual reform, whose alchemy and prophecy were applied to purposes far beyond what he had imagined and whose quintessence ultimately undermined the most basic assumptions of the late medieval world that he knew. The history of John of Rupescissa and of the quintessence—like all histories—is thus one of change, the very coming into being and passing away that Rupescissa tried so hard to suppress.

NOTES

1. INTRODUCTION

1. The standard study of Rupescissa is Jeanne Bignami-Odier, *Études sur Jean de Roquetaillade (Johannes de Rupescissa)* (Paris: Vrin, 1952), based on Bignami-Odier's École des Chartes thesis of 1925. It appears in a revised form (with slightly updated bibliography and some abridgement of the original text) as "Jean de Roquetaillade (de Rupescissa), Théologien, Polémiste, Alchimiste," *Histoire littéraire de la France* 41 (1981): 75–240. My citations are to the revised article. Robert E. Lerner expands upon Bignami-Odier's work in his thorough and learned "Historical Introduction," in *Liber secretorum eventuum* by John of Rupecissa, ed. Robert E. Lerner and Christine Morerod-Fattebert (Fribourg: Editions Universitaires, 1994), 13–85.

2. A critical edition has recently been published as Jean de Roquetaillade (John of Rupescissa), *Liber ostensor quod adesse festinant tempora*, ed. Clémence Thévenaz Modestin, Christine Morerod-Fattebert, André Vauchez, et al. (Rome: École Française de Rome, 2005) (cited hereafter as LO).

3. LO, 516: "dealbari ut in eo posset pingi clara intelligencia revelandi archani."

4. LO, 516: "qui hec debebant secreta capere, oportebat eos decoqui in igne tribulationum multarum. Ideo dicit: *et quasi ignis probabuntur* (Dn 12,10). XII enim annis jam fere coquor in carceribus, et dequoctiones igneales non cessant, sed fortius contra me continue inardescunt.... Et ideo placuit Deo mihi *fantastico* et deliro revelare secretum."

5. Throughout this book, I use the word "naturalism" to describe the systematic study of nature through disciplines such as natural philosophy, astronomy/astrology, medicine, and alchemy. These disciplines, along with other areas of study, make up the body of what we might call "science" in the Middle Ages, but the term is generally viewed as anachronistic when applied to the premodern period.

6. See Robert P. Multhauf, "John of Rupescissa and the Origin of Medical Chemistry," *Isis* 45 (1954): 359–67; Robert Halleux, "Les ouvrages alchimiques de Jean de Roquetaillade," *Histoire littéraire de la France* 41 (1981): 241–84. See also John of Rupescissa, *Johannes' de Rupescissa "Liber de consideratione quintae essentiae omnium rerum" deutsch: Studien*

I. INTRODUCTION

zur Alchemia medica des 15. bis 17. Jahrhunderts mit kritischer Edition des Textes, ed. Udo Benzenhöfer (Stuttgart: Franz Steiner, 1989), 55–82; Giancarlo Zanier, "Procedimenti farmacologici e pratiche chemioterapeutiche nel *De consideratione quintae essentiae*," in *Alchimia e medicina nel medioevo*, ed. Chiara Crisciani and Agostino Paravicini Bagliani (Florence: Sismel, 2003), 161–76.

7. Multhauf, "Origin," 359–60, 366.

8. Multhauf, "Origin"; Multhauf, "The Science of Matter," in *Science in the Middle Ages*, ed. David C. Lindberg (Chicago: University of Chicago Press, 1978), 379. See also Halleux, "Les ouvrages alchimiques," 274–76. Lynn Thorndike offers a useful portrait of Rupescissa in *A History of Magic and Experimental Science*, 8 vols. (New York: Columbia University Press, 1923–58), 3:347–69.

9. Important critical editions, translations, and studies include: Petrus Bonus of Ferrara (Pietro Bono da Ferrara), *Preziosa margarita novella: Edizione del volgarizzamento, introduzione e note*, ed. Chiara Crisciani (Florence: Nuova Italia Editrice, 1976); Chiara Crisciani, *Il Papa e l'alchimia: Felice V, Guglielmo Fabri e l'elixir* (Rome: Viella, 2002); Ramón Llull, *Il "Testamentum" alchemico attribuito a Raimondo Lullo: Edizione del testo latino e catalano dal manoscritto Oxford, Corpus Christi College, 244*, ed. Michela Pereira and Barbara Spaggiari (Florence: Sismel, 1999); Michela Pereira, *The Alchemical Corpus Attributed to Raymund Llull* (London: Warburg Institute, 1989); Pseudo-Geber, *The Summa Perfectionis of Pseudo-Geber: A Critical Edition, Translation and Study*, ed. William R. Newman (Leiden: Brill, 1991); Constantine of Pisa, *The Book of the Secrets of Alchemy*, ed. Barbara Obrist (Leiden: Brill, 1990). See also (Pseudo-)Arnald of Vilanova, *Le Rosier alchimique de Montpellier (Lo Rosari): Textes, traduction, notes, et commentaires*, ed. Antoine Calvet (Paris: Presses de l'Université de Paris-Sorbonne, 1997).

10. Claude Gagnon, "Alchimie, techniques, et technologie," in *Les arts mécaniques au moyen âge, Cahiers d'études médiévales*, vol. 7 (Paris: Vrin, 1982), 131–46; William R. Newman, *Promethean Ambitions: Alchemy and the Quest to Perfect Nature* (Chicago: University of Chicago Press, 2004); Michela Pereira, "Alchemy and the Use of Vernacular Languages in the Late Middle Ages," *Speculum* 74 (1999): 336–56.

11. Betty Jo Teeter Dobbs, *The Janus Faces of Genius: The Role of Alchemy in Newton's Thought* (Cambridge: Cambridge University Press, 1991); Dobbs, *The Foundations of Newton's Alchemy; or, "The Hunting of the Greene Lyon"* (Cambridge: Cambridge University Press, 1975); Lawrence M. Principe, *The Aspiring Adept: Robert Boyle and His Alchemical Quest, Including Boyle's "Lost" Dialogue on the Transmutation of Metals* (Princeton, N.J.: Princeton University Press, 1998).

12. Michela Pereira and Chiara Crisciani offer some valuable insights into alchemy and religion, but they treat Rupescissa's writings only briefly. See their *L'arte del sole e della luna: alchimia e filosofia nel medioevo* (Spoleto: Centro italiano di studi sull'alto medioevo, 1996), 73–74, 219–25; and "Black Death and Golden Remedies: Some Remarks on Alchemy and the Plague," in *The Regulation of Evil: Social and Cultural Attitudes to Epidemics in the Late Middle Ages*, ed. Agostino Paravicini Bagliani and Francesco Santi (Florence: Sismel, 1998), 13–22. See also Johannes Fried, *Aufstieg aus dem Untergang: apokalyptisches Denken und*

die Entstehung der modernen Naturwissenschaft im Mittelalter (Munich: C. H. Beck, 2001). Fried argues that apocalypticism played a role in the growth of experimental research and the Scientific Revolution, but again Rupescissa appears only briefly in his discussion.

13. Despite the title of Bignami-Odier's article, "Jean de Roquetaillade (de Rupescissa), Théologien, Polémiste, Alchimiste" (see note 1), Rupescissa's alchemy is not discussed. See also Marjorie Reeves, *The Influence of Prophecy in the Later Middle Ages: A Study in Joachimism* (Oxford: Clarendon Press, 1969); Robert E. Lerner, *The Feast of Saint Abraham: Medieval Millenarians and the Jews* (Philadelphia: University of Pennsylvania Press, 2000), 73–83; Lerner, "'Popular Justice': Rupescissa in Hussite Bohemia," in *Eschatologie und Hussitismus*, ed. Alexander Patschovsky, František Šmahel, and Antonín Hrubý (Prague: Historisches Institut, 1996), 39–42, among other works. The approach of these studies is shared by the essays on Rupescissa collected in André Vauchez, ed., *Textes prophétiques et la prophétie en Occident (XIIe-XVIe siècle)* (Rome: École Française de Rome, 1990). Several other studies have appeared, including Marc Boilloux, "Étude d'une commentaire prophétique de XIV siècle: Jean de Roquetaillade et l'*Oracle Cyrille* (v. 1345–1349)," dissertation, École des Chartres, 1993; Josep Perarnau i Espelt, "La traducció catalana resumida del *Vade mecum in tribulatione* (Ve ab mi en Tribulació) de Fra Joan de Rocatalhada," *Arxiu de Textos Catalans Antics* 12 (1993): 43–140; Perarnau i Espelt, "La traducció catalana medieval del *Liber secretorum eventuum* de Joan de Rocatalhada," *Arxiu de Textos Catalans Antics* 17 (1998): 7–219; Elizabeth Casteen, "John of Rupescissa's Letter *Reverendissime pater* (1350) in the Aftermath of the Black Death," *Franciscana* 6 (2004): 139–84.

14. Joseph Ziegler, *Medicine and Religion, c. 1300: The Case of Arnau of Vilanova* (Oxford: Clarendon Press, 1998), uses a similar model. He treats the spiritual and naturalist writings of Arnald of Vilanova, whom I discuss in chapter 5.

15. See, for instance, Theodore L. Brown, *Making Truth: Metaphor in Science* (Urbana: University of Illinois Press, 2003); Andrew Ortony, ed., *Metaphor and Thought* (Cambridge: Cambridge University Press, 1993); George Lakoff and Mark Johnson, *Metaphors We Live By* (Chicago: University of Chicago Press, 1980).

16. Halleux lists these manuscripts in "Ouvrages," 278–84. Udo Benzenhöfer and Michela Pereira have discovered additional copies: see, for instance, John of Rupescissa, *Johannes' de Rupescissa*, ed. Benzenhöfer, 26–42, Pereira, "Alchemy and the Use of Vernacular Languages," 345, 350n. 80. See also Andrea Aromatico, "Premessa al testo," in *Il libro della luce*, ed. Andrea Aromatico (Marsilio: Editori, 1997), 116. For more on the print history of Rupescissa's texts, see Rudolf Hirsch, "The Invention of Printing and the Diffusion of Alchemical and Chemical Knowledge," in *The Printed Word: Its Impact and Diffusion* (London: Variorum, 1978), 10:115–41.

2. THE PROVING OF CHRISTENDOM

1. John of Rupescissa, *Vade mecum in tribulatione* (hereafter VM), int. 5: "antequam perveniamus ad annum domini millesimum ccclxv apparebit publice orientalis antichristus

cuius discipuli in partibus yerosolomitanis publice praedicabunt cum falsis signis et portentis in omnem seductionem erroris tertio infra illos prefatos annos v scilicet ab anno domini m ccclx usque [...] v ipsi dabunt [...] clades ultra omnem extimationem humanam tempestates de celis alias nunquam vise diluvia aquatica inaudita in multis partibus orbis praeter diluvium generale fames gravissime supra mundum pestientiales mortalitates et gutturum sanguinatione et alie apostematice passiones quibus plagis interficietur maxima pars grav[is] generationis presentis." A somewhat unsatisfactory early modern edition of the *Vade mecum* appears in *Appendix ad fasciculum rerum expetendarum et fugiendarum*, ed. Edward Brown (London: Richard Chiswell, 1690), 2:496–508. I have studied four manuscripts of this work: Tours, Bibliothèque municipale, MS 520, fol. 32v–47v; Vatican City, Biblioteca Apostolica Vaticana, Lat. 4265, fol. 175–81v; Vatican City, Biblioteca Apostolica Vaticana, Reg. Lat. 1964, fol. 196–203v; Venice, Biblioteca Nazionale Marciana, MS Lat. III, 39, fol. 1–19v. For this study, I cite from the version found in the Venice manuscript, which has been dated to 1359. Of the copies that I have examined, this version provides the best reading, and Robert E. Lerner has identified it as an exemplar of the second recension. I cite here by "intention" (that is, Rupescissa's version of chapters) so that my citations can be found in Brown's edition.

2. On last things, see Caroline Walker Bynum and Paul Freedman, *Last Things: Death and the Apocalypse in the Middle Ages* (Philadelphia: University of Pennsylvania Press, 2000), 1–10.

3. With these definitions, I follow Stephen D. O'Leary, *Arguing the Apocalypse: A Theory of Millennial Rhetoric* (New York: Oxford University Press, 1994), 5–6; Robert E. Lerner, "The Black Death and Western Eschatological Mentalities," *American Historical Review* 86 (1981): 537–38; Richard Landes, "Lest the Millennium Be Fulfilled: Apocalyptic Expectations and the Pattern of Western Chronography 100–800," in *The Use and Abuse of Eschatology in the Middle Ages*, ed. Werner Verbeke, Daniel Verhelst, and Andries Welkenhuysen (Leuven: Leuven University Press, 1988), 205–8.

4. On apocalypse defined as a literary genre, see the work of John J. Collins, "Introduction: Towards the Morphology of a Genre," and Adela Yarbro Collins, "The Early Christian Apocalypses," both in *Semeia* 14 (1979), 1–20, 61–121. See also John J. Collins, *The Apocalyptic Imagination: An Introduction to Jewish Apocalyptic Literature* (1984; reprint, Grand Rapids, Mich.: Eerdmans, 1998), 1–42. For apocalyptic eschatology and spirituality, see Bernard McGinn, *Apocalyptic Spirituality* (New York: Paulist Press, 1979); McGinn, "Introduction: John's Apocalypse and the Apocalyptic Mentality," in *The Apocalypse in the Middle Ages*, ed. Bernard McGinn and Richard K. Emmerson (Ithaca, N.Y.: Cornell University Press, 1992), 5.

5. Joseph F. Zygmunt, "Prophetic Failure and Chiliastic Identity: The Case of Jehovah's Witnesses," in *Expecting Armageddon: Essential Readings in Failed Prophecy*, ed. Jon R. Stone (New York: Routledge, 2000), 65.

6. Lerner, "The Black Death," 552.

7. Rupescissa cited both Sibylline and Joachite prophecies in Jean de Roquetaillade (John of Rupescissa), *Liber ostensor quod adesse festinant tempora*, ed. Clémence Thévenaz

2. THE PROVING OF CHRISTENDOM

Modestin, Christine Morerod-Fattebert, André Vauchez, et al. (Rome: École Française de Rome, 2005) (cited hereafter as LO). There is also evidence that Rupescissa knew Joachim's genuine *Liber concordiae*. See LO, 767.

8. Frank Kermode, *The Sense of an Ending: Studies in the Theory of Fiction* (Oxford: Oxford University Press, 1966).

9. E. Ann Matter, "The Apocalypse in Early Medieval Exegesis," in *The Apocalypse in the Middle Ages*, ed. Bernard McGinn and Richard K. Emmerson (Ithaca, N.Y.: Cornell University Press, 1992), 38–50.

10. Augustine, *De civitate Dei* (Cambridge: Harvard University Press, 1957–72), 6:78–81. "Frustra igitur annos, qui remanent huic saeculo computare ac definire conamur, cum hoc scire non esse nostrum ex ore Veritatis audiamus; quos tamen alii quadringentos, alii quingentos, alii etiam mille ab ascensione Domini usque ad eius ultimum adventum compleri posse dixerunt. Quem ad modum autem quisque eorum astruat opinionem suam, logum est demonstrare et non necessarium. Coniecturis quippe utuntur humanis, non ab eis aliquid certum de scripturae canonicae auctoritate profertur. Omnium vero de hac re calculantium digitos resolvit et quiescere iubet ille qui dicit: *Non est vestrum scire tempora, quae Pater posuit in sua potestate.*"

11. See Bernard McGinn, *Apocalyptic Spirituality*, xv; McGinn, "Early Apocalypticism: The Ongoing Debate," in *Apocalypticism in the Western Tradition* (Aldershot: Variorum, 1994), 1:28; Robert E. Lerner, "Millennialism," in *Apocalypticism in Western History and Culture*, ed. Bernard McGinn, vol. 2 of *The Encyclopedia of Apocalypticism* (New York: Continuum, 2000), 329.

12. The "Tiburtine Sibyl" was written in the Christian East c. 380 and appeared in Latin versions c. 1000. It was extremely popular and frequently revised. The Latin text of Pseudo-Methodius appears in *Die Apokalypse des Pseudo-Methodius, die ältesten griechischen und lateinischen Übersetzungen*, ed. W. J. Aerts and G. A. A. Kortekaas (Louvain: Peeters, 1998). See also Anke Holdenried, *The Sibyl and Her Scribes: Manuscripts and Interpretation of the Latin* Sibylla Tiburtina, *c. 1050–1500* (Aldershot: Ashgate, 2006); Paul Alexander, "Medieval Apocalypses as Historical Sources," *American Historical Review* 73 (1968): 997–1018. Translated excerpts of the Sibylline texts are available in Bernard McGinn, ed., *Visions of the End: Apocalyptic Traditions in the Middle Ages* (1979; New York: Columbia University Press, 1998), 43–50, 70–76.

13. See Marjorie Reeves, *The Influence of Prophecy in the Later Middle Ages: A Study in Joachimism* (Oxford: Clarendon Press, 1969), 299–301.

14. Lerner, "Millennialism," 329–32.

15. Bernard McGinn makes this helpful contrast in his "Joachim of Fiore's *Tertius Status*: Some Theological Appraisals," in *Apocalypticism in the Western Tradition* (Aldershot: Variorum, 1994), 10:220.

16. There is a large body of work on Joachim of Fiore's prophecy and legacy. For an overview, see Marjorie Reeves, *The Influence of Prophecy*; Reeves, *Joachim of Fiore and the Prophetic Future: A Medieval Study in Historical Thinking* (London: S.P.C.K., 1976; reprint, Stroud: Sutton, 1999); Marjorie Reeves and Beatrice Hirsch-Reich, *The Figurae of Joachim of Fiore*

2. THE PROVING OF CHRISTENDOM

(Oxford: Clarendon Press, 1972); Bernard McGinn, *The Calabrian Abbot: Joachim of Fiore in the History of Western Thought* (New York: Macmillan, 1985); Gian Luca Potestà, *Il tempo dell'Apocalisse: Vita di Gioacchino da Fiore* (Rome: Laterza, 2004).

17. Reeves, *Prophetic Future*, 29–30.

18. On the innovative nature of these *figurae*, see Marjorie Reeves, "Joachim of Fiore and the Images of the Apocalypse According to St. John," *Journal of the Warburg and Courtauld Institutes* 64 (2001): 281–95.

19. See Robert E. Lerner, "Refreshment of the Saints: the Time After Antichrist as a Station for Earthly Progress in Medieval Thought," *Traditio* 32 (1976): 97–144; on Antichrist, see Lerner, "Antichrists and Antichrist in Joachim of Fiore," *Speculum* 60 (1985): 553–70.

20. Harold Lee, Marjorie Reeves, and Giulio Silano, *Western Mediterranean Prophecy: The School of Joachim of Fiore and the Fourteenth-Century Breviloquium* (Toronto: Pontifical Institute of Mediaeval Studies, 1989), 3. On the role of Frederick II in the Joachite schema, see Robert E. Lerner, "Frederick II, Alive, Aloft, and Allayed in Franciscan-Joachite Eschatology," in *The Use and Abuse of Eschatology in the Middle Ages*, ed. Werner Verbeke, Daniel Verhelst, and Andries Welkenhuysen (Leuven: Leuven University Press, 1988), 359–84.

21. For an introduction to St. Francis and the origins of his order, see John Moorman, *A History of the Franciscan Order: From Its Origins to the Year 1517* (Oxford: Clarendon Press, 1968); Lester K. Little, *Religious Poverty and the Profit Economy in Medieval Europe* (Ithaca, N.Y.: Cornell University Press, 1978); Cajetan Esser, O.F.M., *Origins of the Franciscan Order*, trans. Aedan Daly and Irina Lynch (Chicago: Franciscan Herald Press, 1970); Rosalind B. Brooke, *The Coming of the Friars* (London: George Allen and Unwin, 1975).

22. David Burr, *Olivi and Franciscan Poverty: The Origins of the Usus Pauper Controversy* (Philadelphia: University of Pennsylvania Press, 1989), 4.

23. Moorman, *A History of the Franciscan Order*, 90–91, 117.

24. The standard work on the Franciscan controversy is David Burr, *The Spiritual Franciscans* (University Park: Pennsylvania State University Press, 2001). See also Malcolm Lambert, *Franciscan Poverty: The Doctrine of the Absolute Poverty of Christ in the Franciscan Order, 1210–1323* (London: S.P.C.K., 1961); Little, *Religious Poverty and the Profit Economy*; and the essays collected in *Franciscains d'Oc: Les spirituels ca. 1280–1324*, Cahiers de Fanjeaux 10 (1975). A number of general studies also discuss the Spirituals. See, for instance, Malcolm Lambert, *Medieval Heresy: Popular Movements from the Gregorian Reform to the Reformation*, 3rd ed. (Oxford: Blackwell, 2002), 208–35; Gordon Leff, *Heresy in the Later Middle Ages: The Relation of Heterodoxy to Dissent c. 1250–1450*, 2 vols. (Manchester: Manchester University Press, 1967), 1:51–255. There is also an excellent survey of the Beguins in southern France: Louisa A. Burnham, *So Great a Light, So Great a Smoke: The Beguin Heretics of Languedoc* (Ithaca, N.Y.: Cornell University Press, 2008).

25. On Olivi, see David Burr, *Olivi and Franciscan Poverty*; Burr, *Olivi's Peaceable Kingdom: A Reading of the Apocalypse Commentary* (Philadelphia: University of Pennsylvania Press, 1993); Burr, *The Persecution of Peter Olivi*, Transactions of the American Philosophical Society 66,

no. 5 (Philadelphia: American Philosophical Society, 1976); Burr, "Bonaventure, Olivi, and Franciscan Eschatology," *Collectanea Franciscana* 53 (1983): 23–40; Burr, "Olivi's Apocalyptic Timetable," *Journal of Medieval and Renaissance Studies* 11 (1981): 237–60; and Burr, "Olivi, Apocalyptic Expectation, and Visionary Experience," *Traditio* 41 (1985): 273–88.

26. An edition of Olivi's Apocalypse commentary appears in Warren Lewis's unpublished but widely read edition of the *Lectura super Apocalipsim*, a part of his "Peter John Olivi: Prophet of the Year 2000. Ecclesiology and Eschatology in the *Lectura Super Apocalipsim*," 2 vols., Ph.d. diss., Tübingen, 1972.

27. Robert E. Lerner, "Writing and Resistance Among Beguins of Languedoc and Catalonia," in *Heresy and Literacy, 1000–1530*, ed. Peter Biller and Anne Hudson (Cambridge: Cambridge University Press, 1994), 188–89; Leff, *Heresy in the Later Middle Ages*, 161.

28. Burr, *Olivi's Peaceable Kingdom*, 191–92.

29. See Burr's discussion in *Olivi's Peaceable Kingdom*, ix–xii, 70. See also Lewis, "Peter John Olivi: Author of the *Lectura super apocalipsim*: Was He Heretical?" in *Pierre de Jean Olivi (1248–1298): Pensée scolastique, dissidence spirituelle et société: actes du colloque de Narbonne, mars 1998*, ed. Alain Boureau and Sylvain Piron (Paris: Vrin, 1999), 135–56.

30. Reeves, *Influence of Prophecy*, 62. See also Annie Cazenave, "La vision eschatologique des spirituels franciscains autour de leur condamnation," in *The Use and Abuse of Eschatology in the Middle Ages*, ed. Werner Verbeke, Daniel Verhelst, and Andries Welkenhuysen (Leuven: Leuven University Press, 1988), 393–403. Joachim's trinitarian doctrine had already been taken by Gerard of Borgo San Donnino in 1254 as evidence that the authority of the institutional church would be completely abrogated as the states of the Father and Son were eclipsed by the state of the Holy Spirit.

31. Reeves, *Influence of Prophecy*, 175.

32. Cited from John XXII's bull, *Gloriosam ecclesiam*, in Burr, *Spiritual Franciscans*, 200.

33. See Reeves, *Influence of Prophecy*, 69.

34. Burr and Lerner both report this observation of Angelo Clareno; for his remarks, see Angelo of Clareno, *Epistole*, vol. 1 of *Angeli Clareni Opera*, ed. Lydia Von Auw (Rome: Nella sede dell'Istituto, 1980), 174–75.

35. My summary draws upon Lerner, "Writing and Resistance," 186–204; Leff, *Heresy in the Later Middle Ages*, 195–255; Burnham, *So Great a Light*.

36. Bernard Gui, *Manuel de l'inquisiteur*, ed. G. Mollat, 2 vols. (Paris: Les Belles Lettres, 1877–1968), 1:144–46. The Beguins predicted that the Franciscan order would be split into three groups: the Conventual Franciscans, the Fraticelli (Franciscan dissenters in Italy), and the true Franciscans and their followers. During the time of Antichrist, the former two would be destroyed; the latter third would survive because of their adherence to the rigors of the rule.

37. On this date, see Josep Perarnau i Espelt, "El text primitiu del *De mysterio cymbalorum ecclesiae* d'Arnau de Vilanova," *Arxiu de Textos Catalans Antics* 7/8 (1988–89): 92, note on line 727.

2. THE PROVING OF CHRISTENDOM

38. Burr, *Spiritual Franciscans*, 97–98. See also Burr, *Olivi and Franciscan Poverty*, 182–83; as he notes, Olivi made identifications of true persons with positive apocalyptic characters, such as Francis with the angel of the apocalypse.

39. Burr, *Olivi's Peaceable Kingdom*, 71.

40. Lerner, "Writing and Resistance," 197.

41. Burr, *Spiritual Franciscans*, 196–206.

42. Rupescissa expressed a great deal of sympathy for these burned Franciscans and Beguins, calling them in his *Liber ostensor* martyrs burned for evangelical poverty; their Dominican persecutors were the "heretics of Mammon" (*heretici mammonisti*). See, for instance, LO, 229–32, 525, and passim.

43. For *Genus nequam* and *Ascende calve*, see Orit Schwartz and Robert E. Lerner, "Illuminated Propaganda: The Origins of the 'Ascende Calve' Pope Prophecies," *Journal of Medieval History* 20 (1994): 157–91. Rupescissa refers to both illustrated prophecies in LO (303, 347, 359, 402, 411, etc.). His full-length commentary on *Ascende calve* is no longer extant. On illuminated pope prophecies, see also Hélène Millet, *Les successeurs du pape aux ours: Histoire d'un livre prophétique médiéval illustré* (Turnhout: Brepols, 2004). For other prophecies cited by Rupescissa, see LO, 865–956.

44. Lerner, "Historical Introduction," in *Liber secretorum eventuum* by John of Rupescissa, ed. Robert E. Lerner and Christine Morerod-Fattebert (Fribourg: Editions Universitaires, 1994), 20–21.

45. LO, 326, 372.

46. John of Rupescissa, *Vos misistis*, in *Appendix ad fasciculum rerum expetendarum et fugiendarum*, ed. Edward Brown (London: Richard Chiswell, 1690), 2:494. The letter is reportedly a response to the archbishop of Toulouse's question: "How long will the wars which have been and still are in France endure?" An English translation appears in Jean de Venette, *The Chronicle of Jean de Venette*, trans. Jean Birdsall, ed. Richard A. Newhall (New York: Columbia University Press, 1953), 61–62.

47. Lerner, "Historical Introduction, 20–25.

48. It is possible that Rupescissa's account of being tossed in the mud is figurative, but his precise language, "in actual soft mud" (*in vero luto molli*), leads me to believe his description is literal. See LO, 517.

49. LO, 517–18: "Bis et tertia vice fui positus in torculari ad tibiam tortam retificandum et tractus usque ad ultimos cruciatus, eo modo quo martirum antiquitus trahebantur membra usque ad extractionem fere; preter hoc quod fere centum diebus, stans in grabato, non valens me movere non visitabar, nisi bis vel ter in die, quando flebilia victualia mihi frater unus portabat et me statim solum relinquebat clausum ut canem, et subtus me nature necessaria faciebam, tanta abilitate quanta potest se levare qui tibiam habet fractam; preter hoc quod de LX diebus mihi semel stramentum mor[t]iferum preparatum non fuit, cum hoc quod tanta multitudo vermium scaturiebat in tibia fracta quod cum duabus manibus de corpore meo colligebantur et exterius portabantur."

50. LO, 519: "infirmum, debilem et flentem."

51. LO, 518–20.

2. THE PROVING OF CHRISTENDOM

52. LO, 518–20.

53. LO, 520: "fui auxilio et servicio destitutus, ita ut de tota die non esset aliquando qui michi ciphum aque ministraret aut qui aspiceret an essem adhuc vivus."

54. See the contemporary Latin *procès-verbal* of the trial and a later German version, both edited in Jeanne Bignami-Odier, "Jean de Roquetaillade (de Rupescissa), Théologien, Polémiste, Alchimiste," *Histoire littéraire de la France* 41 (1981): 214–18. A similar account of the trial appears in the chronicle of Henry of Herford; see Bignami-Odier, "Jean de Roquetaillade," 212.

55. Lerner, "Historical Introduction," 30.

56. LO, 524: "firmiter credam quod omnes sancti qui fuerunt ab origine mundi non sustinuerunt tot verbales injurias quot ego solus passus sum nocte et die in quinque annis a demoniaco illo."

57. LO, 524: "Et quia novum est valde, ingeniavit michi novum genus dilapidationis, quoniam, habens cabacium quemdam, implebit illud de stercoribus suis. Locus quidem, ubi ego studebam habebat minus postem unum juxta me, per quem locum apertum me lapidibus obruebat interdum. Caute autem quadam die, cum sederem ad studium, proiecit cabacium plenum stercoribus contra parietem, et reverberando super me stercora ceciderunt."

58. LO, 524: "Anno siquidem Domini mccclv vel circa, in estate, quodam vespere, cum misisset michi Deus bonam cibariorum elemosinam et partem darem cuidam innocenti iniuste incarcerato ad parvulum portanellum ubi incarceratis cibaria traduntur, abscondit se dictus demoniacus retro portanellum. Fecerat enim quamdam brocam osseam acutissimam sicut stilus ferreus; et, cum infra portanellum posuissem manum cum cibariis porri–gendo pauperi carcerato, prefatus proditor anglicus, latenter irruens in me, cum una manu arripuit me per capicium capucii et cum alia manu in qua brocam acutam tenebat percussiens immanissimo impetu contra pectus meum, me interficiebat, nisi quia incarceratus cui elemosinam dabam manum suam ictui mortifero opposuit, et jussu Christi me liberavit a morte. Manus tamen sua perforata remansit et cicatrizata usque in hodernum diem."

59. There is some uncertainty about the date of the *Liber lucis*. Lynn Thorndike gives several possible dates—ranging from 1350 to 1380—following the dates given in various manuscript copies, although dates after 1365 are certainly incorrect. Robert Halleux has found manuscripts dating the treatise to 1330, 1350, and 1354. A date during the 1350s, when Rupescissa was imprisoned in Avignon, seems most probable. See Thorndike, *A History of Magic and Experimental Science*, 8 vols. (New York: Columbia University Press, 1923–58), 3:352; Halleux, "Les ouvrages alchimiques de Jean de Roquetaillade," *Histoire littéraire de la France* 41 (1981): 267.

60. See, for instance, LO, 327. On the growth of the papal library, see Marie-Henriette Jullien de Pommerol, "Les papes d'Avignon et leurs manuscrits," in *Livres et bibliothèques (XIIIe–XVe siècle), Cahiers de Fanjeaux* 31 (1996): 133–56. Sylvie Barnay has argued that Rupescissa was far less radical a Franciscan than many of his predecessors, including Angelo Clareno, Ubertino of Casale, and even Peter Olivi. She speculates that he

2. THE PROVING OF CHRISTENDOM

became entangled in a power struggle between Elias of Talleyrand-Périgord, who was his protector, and Gui of Bologna, who had his own rival prophet. See her "L'univers visionnaire de Jean de Roquetaillade," in *Fin du monde et signes des temps: Visionnaires et prophètes en France méridionale (fin XIIIe-début XVe siècle)*, *Cahiers de Fanjeaux* 27 (1992): 171–90.

61. John of Rupescissa, *Reverendissime pater*, in *Appendix ad fasciculum rerum expetendarum et fugiendarum*, ed. Edward Brown (London: Richard Chiswell, 1690), 2:494–96, and *Vos misistis*, 2:494. On the former, see Elizabeth Casteen, "John of Rupescissa's Letter *Reverendissime pater* (1350) in the Aftermath of the Black Death," *Franciscana* 6 (2004): 139–84.

62. LO, 143: "et ecce una persona que me diligit prophetiam michi promissam ad litteram misit. Quam cum legissem, stupens quod, sicut domina michi dixerat, ibidem continerentur prophetie, que se habent ad Antichristum et tribulationem propinquam."

63. VM, prologue.

64. Lerner, "Historical Introduction," 83–84.

65. See, for instance, Lerner's account of the rapid translation of the *Vade mecum* into French in his "Medieval Millenarianism and Violence," in *Pace e guerra nel basso medioevo: Atti del XL convegno storico internazionale* (Spoleto: Centro italiano di studi sull'alto medioevo, 2004), 37–52. There is no comprehensive list of vernacular translations of Rupescissa's works. Bignami-Odier, "Jean de Roquetaillade," lists a number of them in her bibliography, 222–40. Lerner notes Italian and Bohemian versions of the *Vade mecum* in his "'Popular Justice': Rupescissa in Hussite Bohemia," in *Eschatologie und Hussitismus*, ed. Alexander Patschovsky, František Šmahel, and Antonín Hrubý (Prague: Historisches Institut, 1996), 42, 50–51. Josep Perarnau i Espelt provides an edition of a Catalan version of the *Vade mecum* in his "La traducció catalana resumida del *Vade mecum in tribulatione* (Ve ab mi en Tribulació) de Fra Joan de Rocatalhada," *Arxiu de Textos Catalans Antics* 12 (1993), and of the *Liber secretorum eventuum* in his "La traducció catalana medieval del *Liber secretorum eventuum* de Joan de Rocatalhada," *Arxiu de Textos Catalans Antics* 17 (1998). A German version of *De quinta essentia* (and a survey of its influence on German alchemy) appears in John of Rupescissa, *Johannes' de Rupescissa "Liber de consideratione quintae essentiae omnium rerum" deutsch: Studien zur Alchemia medica des 15. bis 17. Jahrhunderts mit kritischer Edition des Textes*, ed. Udo Benzenhöfer (Stuttgart: Franz Steiner, 1989). Marguerite Ann Halversen provides an edition of a Middle English *De quinta essentia* in her "'The Consideration of Quintessence': An Edition of a Middle English Translation of John of Rupescissa's *Liber de Consideratione de Quintae Essentiae Omnium Rerum* with Introduction, Notes, and Commentary," Ph.D. diss., Michigan State University, 1998.

66. Jean de Venette, *Chronicle*, 61 (year 1356).

67. Jean de Venette, *Chronicle*, 213. Newhall identifies the *Chronicon Moguntinum*'s unnamed friar with Rupescissa. Bignami-Odier, "Jean de Roquetaillade," 211, notes that it was actually Clement VI who imprisoned Rupescissa, but that he remained captive during the pontificate of Clement's successor.

68. Much has been written about the disasters of the fourteenth century. See, for instance, C. T. Allmand, *The Hundred Years War: England and France at War, c. 1300–c. 1450*

(Cambridge: Cambridge University Press, 1988); William Chester Jordan, *The Great Famine: Northern Europe in the Early Fourteenth Century* (Princeton, N.J.: Princeton University Press, 1996); Rodney Hilton, *Bond Men Made Free: Medieval Peasant Movements and the English Rising of 1381* (New York: Viking Press, 1973); Philip Ziegler, *The Black Death* (London: Collins, 1969); Samuel K. Cohn, *The Black Death Transformed: Disease and Culture in Early Renaissance Europe* (New York: Oxford University Press, 2002); Jean Noël Biraben, *Les hommes et la peste en France et dans les pays européens et méditerranéens*, 2 vols. (Paris: Mouton, 1975–1976); Harry A. Miskimin, *The Economy of Early Renaissance Europe, 1300–1460* (Englewood Cliffs, N.J.: Prentice-Hall, 1969).

69. On the Avignon papacy, see G. Mollat, *The Popes at Avignon, 1305–1378* (New York: Harper and Row, 1965).

70. VM, int. 4: "Intentio quarta est certissima quota anni et diei infra quem fugiet curia romana de civitate peccatrice avinione firmiter teneatis in tantum tribulationis expresse subito in universo mundo ut anno domini mccclxii et xv die julii fugient domini cardinales de amena requie avinionis huius a facie tribulationis et turbationis."

3. JOHN OF RUPESCISSA'S VISION OF THE END

1. Jean de Roquetaillade (John of Rupescissa), *Liber ostensor quod adesse festinant tempora*, ed. Clémence Thévenaz Modestin, Christine Morerod-Fattebert, André Vauchez, et al. (Rome: École Française de Rome, 2005) (cited hereafter as LO), 530–31: "'Frater Johannes, in libris tuis—quos idem cardinalis tenebat in manu—prophetas quod nos debemus pati tribulationes maximas et humiliari ac perdere divitias et hanc temporalem gloriam quam habemus; et quod debet reverti potestas papalis et actoritas Ecclesie ad quosdam pauperes ordinis tui: que omnia sunt fantastica et delira!'"

2. I cite the revised article, Jeanne Bignami-Odier, "Jean de Roquetaillade (de Rupescissa), Théologien, Polémiste, Alchimiste," *Histoire littéraire de la France* 41 (1981): 198–202. Marjorie Reeves's account of Rupescissa in *The Influence of Prophecy* is indebted to Bignami-Odier, and she similarly connects Rupescissa's writings to French politics and prophecy, including Pierre Dubois's *De recuperatione Terrae Sanctae* and the anonymous *Liber de Flore*, both pro-French tracts. See Reeves, *The Influence of Prophecy in the Later Middle Ages: A Study in Joachimism* (Oxford: Clarendon Press, 1969), esp. 321–24, and *Joachim of Fiore and the Prophetic Future: A Medieval Study in Historical Thinking* (London: S.P.C.K., 1976; reprint, Stroud: Sutton, 1999), 66–69.

3. Among his main works that treat Rupescissa, see Robert E. Lerner, "'Popular Justice': Rupescissa in Hussite Bohemia," in *Eschatologie und Hussitismus*, ed. Alexander Patschovsky, František Šmahel, and Antonín Hrubý (Prague: Historisches Institut, 1996), 39–42; Lerner, "Millennialism," in *Apocalypticism in Western History and Culture*, ed. Bernard McGinn, vol. 2 of *The Encyclopedia of Apocalypticism* (New York: Continuum, 2000), 326–60; Lerner, "Historical Introduction," in *Liber secretorum eventuum* by John of Rupescissa, ed. Robert E. Lerner and Christine Morerod-Fattebert (Fribourg:

Editions Universitaires, 1994), 13–85; and Lerner, "The Black Death and Western Eschatological Mentalities," *American Historical Review* 86 (1981): 533–52.

4. See Robert E. Lerner, "Medieval Millenarianism and Violence," in *Pace e guerra nel basso medioevo: Atti del XL convegno storico internazionale* (Spoleto: Centro italiano di studi sull'alto medioevo, 2004), 48–52.

5. More than forty manuscript copies of the *Vade mecum in tribulatione* are extant. A partial list of manuscripts appears in Bignami-Odier, "Jean de Roquetaillade," 231–37; additional copies and resumes found by Lerner are in "'Popular Justice,'" 50–51. No one has attempted to collect and collate the extant manuscripts of the *Vade mecum*. Lerner has noted that this task will prove particularly challenging because of the large number of variant readings.

6. The *Vade mecum* comprises twenty brief chapters (called *intentiones*) that generally occupy fewer than twenty folio pages, in contrast to the *Liber ostensor*'s 149 and the *Oraculum Cyrilli*'s 244 folio pages. Robert E. Lerner has identified two recensions: one includes twenty intentions, the other seventeen intentions (lacking intentions 17, 18, and 20); both versions are possibly attributable to Rupescissa. See Lerner, "'Popular Justice,'" 39–40n. 2.

7. Bignami-Odier, "Jean de Roquetaillade," 157.

8. Modern scholars have characterized these contrasting attitudes as "negative" and "positive" apocalypticism. For an discussion of this distinction, see Derk Visser, *Apocalypse as Utopian Expectation (800–1500): The Apocalypse Commentary of Berengaudus of Ferrières and the Relationship Between Exegesis, Liturgy, and Iconography* (Leiden: Brill, 1996), 1–2.

9. All of my citations of the *Vade mecum* (VM) are from John of Rupescissa, *Vade mecum in tribulatione*, Venice, Biblioteca Nazionale Marciana, MS Lat. III, 39, fol. 1–19v, but I cite according to "prologue" or "intention" so that readers can find the passages in the published Brown edition. See chapter 2, note 1, for an explanation. VM, prologue: "Gaudeo de reparatione futura sed contristor de imminenti pressura universi populi christiani durissima qualis nunquam fuit ab origine mundi nec postea futura est usque ad finem seculi."

10. VM, int. 20.

11. VM, int. 3.

12. VM, int. 4.

13. VM, int. 3: "Intentio tertia est de gravium futurorum flagellorum ordine et effectum et fine supra enim omnem estimationem humanam pondus instantium flagellorum super clerum vertetur in tantam quod hac vice hoc universa ecclesia nisi quia perire non potest christo domino orante pro ea sed graviter fluctuabit pro certo quia ut multi estiment ipsam penitus suffocatam."

14. VM, int. 5: "per v annos continuos erunt in mundo novitates horrende primo vermes terre tantam fortitudinem et audaciam induentur ut crudelissime devorent fere omnes leones ursos leopardos et lupos et alaude et merulane per uule aves rapaces falchones et accipitres laniabunt hoc est enim necessarium ut impleatur vaticinium ysaie cap 33 vhe qui predaris nonne et ipse predaberis.... Consurget insuper infra illos

quinque annos iusititia popularum et tyrannos proditores nobiles in hore bisacuti gladii devorabit et cadent multi principum et nobilium et potentum a dignitatibus suis et divitiarum suarum et fiet afflictio in nobilibus ultra quam credere possit et rapientur maiores qui cum proditionibus depraedari fecerant populum tam afflictu."

15. Rodney Hilton, *Bond Men Made Free: Medieval Peasant Movements and the English Rising of 1381* (New York: Viking Press, 1973), 96–134; Samuel K. Cohn, *Creating the Florentine State: Peasants and Rebellion, 1348–1434* (Cambridge: Cambridge University Press, 1999); Cohn, *Lust for Liberty: The Politics of Social Revolt in Medieval Europe, 1200–1425: Italy, France, and Flanders* (Cambridge, Mass.: Harvard University Press, 2006).

16. Lerner, "'Popular Justice,'" 42–49.

17. VM, prologue.

18. The idea of multiple Antichrists was common in medieval texts. Joachim of Fiore's figure of the seven-headed dragon suggested a number of Antichrist-persecutors of the church, including a "great" Antichrist and a "last" Antichrist; Olivi similarly accounted for "mystical," "great," and "last" Antichrists in his Apocalypse commentary. See Robert E. Lerner, "Antichrists and Antichrist in Joachim of Fiore," *Speculum* 60 (1985): 553–70; "Historical Introduction," 53; David Burr, *Olivi's Peaceable Kingdom: A Reading of the Apocalypse Commentary* (Philadelphia: University of Pennsylvania Press, 1993), 132–56. Rupescissa proposed three Antichrists in the *Vade mecum*: an eastern Antichrist, a western Antichrist—who was for Rupescissa the great Antichrist—and a final Antichrist, who would be loosed from his chains after the end of the one-thousand-year peace, as described in Apocalypse 20. There has been much scholarly work on perceptions of Antichrist; for an overview, see Richard K. Emmerson, *Antichrist in the Middle Ages: A Study of Medieval Apocalypticism, Art, and Literature* (Seattle: University of Washington Press, 1981); Bernard McGinn, *Antichrist: Two Thousand Years of the Human Fascination with Evil* (San Francisco: Harper, 1994); also see Roberto Rusconi, "Antichrist and Antichrists," in *Apocalypticism in Western History and Culture*, vol. 2 of *The Encyclopedia of Apocalypticism*, ed. Bernard McGinn (New York: Continuum, 2000), 287–325.

19. VM, int. 15.

20. Reeves cites Roger Bacon as the first to mention an Angelic Pope in his *Opus maius*. See *Influence of Prophecy*, 47. See also McGinn, "Angel Pope and Papal Antichrist," and "*Pastor Angelicus*: Apocalyptic Myth and Political Hope in the Fourteenth Century," in *Apocalypticism in the Western Tradition* (Aldershot: Variorum, 1994), 5:155–73; 6:221–51. The Angelic Pope does not appear in either Joachim or Olivi; the idea is thought to have been revitalized by the election of Pope Celestine V in 1294; see Lerner, "Millennialism," 337; "Historical Introduction," 69.

21. VM, int. 12: "hic est ellias qui iuxta verbum christi restituet omnia scilicet corruptos sacerdotes avaros luxuriosos christus cum hoc flagello ad literam facto de funiculis de pauperculis cordellatis abiectis de templo expellet pro certo ne ei ministrent in sacrificio et symoniacos deponet de busso ecclesiastico et offendentes naturam tradat brachio seculi ut sacrificentur in igne et ut purgetur adulterem castigabitur superia cleri ad ultimum in stercorem concludet et libertatem antiquam elligendi praelatos sedibus episcopalibus."

3. JOHN OF RUPESCISSA'S VISION OF THE END

22. Lerner, "Millennialism," 353.

23. Lerner compares Rupescissa's description to the mentions of the millennium that appear in the prophecies of Angelo Clareno, Columbinus, and Ubertino da Casale. See "Historical Introduction," 77. Peter Olivi's description of the third state in his *Lectura* is somewhat more detailed. For an analysis, see Warren Lewis, "Peter John Olivi: Prophet of the Year 2000. Ecclesiology and Eschatology in the *Lectura Super Apocalipsim*," 2 vols., Ph.d. diss., Tübingen, 1972, esp. 1:234–44. For a summary of the *Liber secretorum eventuum*'s millennial predictions, see the second chapter of Michael Alan Ryan's "That the Truth Be Known: Prophecy and Society in the Late Medieval Crown of Aragon," Ph.D. diss., University of Minnesota, 2005, 51–121. For an overview of medieval chiliasm and a discussion of Rupescissa, see Lerner, "The Medieval Return to the Thousand-Year Sabbath," in *The Apocalypse in the Middle Ages*, 51–71.

24. Lerner, "Historical Introduction," 14.

25. Robert E. Lerner, "Refreshment of the Saints: the Time After Antichrist as a Station for Earthly Progress in Medieval Thought," *Traditio* 32 (1976): 102–3, 129.

26. This opinion of Arnald outraged some theologians. John of Paris rejected Arnald's exegetical method and maintained the shorter period of rest; Oxford's Henry of Harclay disputed the application of Ezekiel 4:6's "day for a year" to Daniel. In contrast, the highly regarded exegete Nicholas of Lyra appears to have been swayed by Arnald's new interpretation. See Lerner, "Refreshment," 130–31.

27. Lerner, "Millennialism," 351.

28. Burr, *Olivi's Peaceable Kingdom*, 173–76.

29. John of Rupescissa, *Liber secretorum eventuum*, ed. Robert E. Lerner and Christine Morerod-Fattebert (Fribourg, Switzerland: Editions Universitaires, 1994) (hereafter, LSE), 80: "Evidens enim est quod ibi loquitur de martiribus temporis Antichristi cum dicit: *Et qui non adoraverunt bestiam et imaginem eius* et cet. Isti ergo *vixerunt* corporaliter resuscitati *et regnaverunt* corporaliter *cum Christo mille annis* qui futuri sunt a die mortis Antichristi usque ad adventum Gog et prope finem mundi. Et quia ista resurrectio corporalis non est sanctis omnibus et martiribus aliorum temporum generalis, sed est martiribus temporum Antichristi signanter et privilegialiter specialis, merito subinfertur: *Ceteri mortuorum non vixerunt donec consummentur mille* predicti *anni*, computandi a die mortis Antichristi et ligationis draconis usque ad adventum Gog novissimi et finem huius mundi." This idea also appears in VM, int. 18.

30. LSE, 74: "Fui enim in hac apertione plurimum stupefactus, quia glosis et dictis sanctorum repugnare videtur. Et quia hesitare temptabam—nec mirum!—fuit michi hoc secretum, per me anterius non auditum, datum intelligi contineri in divina Scriptura cuius manifesta probatio est ista."

31. LSE, 84: "Notandam est autem quod ideo conatus fuit Augustinus exponere mistice mille annos predictos quia sibi non fuit datum intelligere quod ligatio draconis predicta deberet fieri in die mortis proximi Antichristi, sicut patet in libro *De Civitate Dei*. Nec in omnibus tenemur sequi expositiones omnium doctorum, quas posuerunt opinando potius quam aliquid contra veritatem textus temere asserendo, nisi ex sacro

3. JOHN OF RUPESCISSA'S VISION OF THE END

textu vel ex determinatione Ecclesie probaretur quod Augustino et ceteris doctoribus olim preteritis fuerunt omnia Scripturarum et prophetiarum archana funditus revelata, ita quod posteris non deberet aliquid ulterius revelari. Cuius oppositum asserit eximius doctor Gregorius omelia nona super Ezechielem, ubi docet posteriores et magis propinquos fini seculi limpidius et clarius intelligere Scripturarum sacrarum veritates occultas."

32. LSE, 86: "Dicent enim heretici novelli futuri quod sancti dixerunt quod statim post mortem Antichristi, datis XLV diebus in penitentia lapsis, veniet finis mundi et resurrectio ab Ecclesia expectata, non intelligentes quod sancti locuti sunt pluries inquirendo et opinando et quod non diffinierunt precise debere post illos XLV dies seculum terminari et resurrectionem generalissimam celebrari, sed dixerunt quod ad minus seculum durabit XLV diebus post mortem Antichristi et amplius quantum volet Deus."

33. One can detect more deference to Augustine in the *Vade mecum*. Rupescissa again calculates the one-thousand-year duration of the millennium according to Apocalypse 20, but he notes that it is not possible to determine the exact date of the eschaton. The one-thousand-year figure is only a minimum.

34. VM, int. 9: "ipse est post christum qui principio tertii status orbis generalis lapis absissus de monte sine manibus qui babillonicam statuam percutiet in pedibus fictilibus."

35. LSE, 102: "Vicesimo quinto intellexi futuros esse post mortem Antichristi XLV annos bellis variis sevissimos in quibus Romanum Imperium occidetur, quia futurum est ut potestas monarchie Romani Imperii in Ierusalem et ultramarinis partibus transmittetur, sicut infra dicetur."

36. LSE, 128: "Sed est pro certo futurum ut, transactis septingentis annis a morte proximi Antichristi, incipiant viri ecclesiastici pro maiori parte prefatam deserere sanctitatem et orbis declinare a puritate iam dicta. Nam multi Ecclesie prelatorum efficientur rapaces avaritia et superbia vite et cultu carnis—supra quam credi potest—et peccatis omnibus supra modum simonia et ambitione infecti; et sub eis peribit iustitia; et innocentes et sancti dabuntur in derisum et graviter affligentur; et incredibilis multitudo Fratrum Minorum efficietur paulatim et insensibiliter a perfectione loginqua."

37. VM, int. 2: "Intentio secunda est universum clerum et dominos supremos ecclesie universe papam et cardinales principatus primates archiepiscopos et episcopos et ceteros reducere ad modum vivendi christi et apostolorum perfectissimum quoniam impossibile foret ab ecclesiae seculam aliter predictam et excecatum reparare quid prodest si enim prelati ecclesie in terris incederent cum CC aut cum CCC equis sicut nonnulli hodierni incedunt et ad praedicationem christi humilitatem cum militibus in pompa tanta ac tali."

38. VM, int. 7: "dabit illis afflictio intellectum et culpas suas humiliter cognoscant et tunc disponent reducere ad modum vivendi christi et apostolorum."

39. He also discusses the divisions of the Franciscans in the LO, although his predictions differ somewhat from the LSE. See, for instance, LO, 621–28.

3. JOHN OF RUPESCISSA'S VISION OF THE END

40. LSE, 28–30.

41. See chapter 2, note 34.

42. This prediction demonstrates an interesting insight into the actual papal schism of 1378; for Rupescissa's reputation as a prophet of the schism, see Robert Swanson, "A Survey of Views on the Great Schism, c. 1395," *Archivum Historiae Pontificiae* 21 (1983): 101–2.

43. VM, int. 14: "sicut expresse dixit dominus beato francischo quod si fratres sui permanerent in observantia regulari christi sicut incepant ostensa beato francischo instans tribulatio non veniret."

44. He addresses both the Michaelists and *Cum inter nonnullos* in LSE, 30–31.

45. VM, int. 14: "Et nisi deus ordini fratrum minorum de praefato reparatore qui eorum ordinem funditus reformabit sicut ceteri remanerent deserti sed de precibus beati francisci et omnium fratrum sanctorum post tribulationem reparabitur ordo et dillatabitur per universum orbem sicut stelle celli que praemultitudine numerari non possunt sed non in superbia habituum et hedificiorum nec in avaritia et ceteris laxationibus sed in modo vivendi christi et apostolorum et sancti francisci et sociorum eiusdem."

46. VM, int. 19; LSE, 125.

47. VM, int. 1: "Intentio prima est in tribulationibus inchoatis et de gradu in gradum agendum totum in unam fidem catholicam ecclesie generalis roman[e] sub obedientia romani unius pontificis summi in uno ovili domenico tantam mundi machinam congregare convertendo iudeos saracenos turchos et tartaras nationes ad veritatem ecclesie generalis et omnes praevos christianos in omnibus mortalibus sceleribus induratos fonditus extirpare de terra." Missionary activity and the conversion of the Muslims was of particular interest to Franciscans; see E. Randolph Daniel, *The Franciscan Concept of Mission in the High Middle Ages* (Lexington: University Press of Kentucky, 1975), esp. 55–74.

48. Robert E. Lerner, *Feast of Saint Abraham: Medieval Millenarians and the Jews* (Philadelphia: University of Pennsylvania Press, 2000), 79; see also his "Millénarisme littéral et vocation des juifs chez Jean de Roquetaillade," in *Textes prophétiques et la prophétie*, ed. André Vauchez (Rome: École Française de Rome, 1990), 21–25.

49. VM, int. 9.

50. LO, 212n. 2; J.-P. Boudet, "La papauté d'Avignon et l'astrologie," in *Fin du monde et signes des tempss: Visionnaires et prophètes en France méridionale fin XIIIe-début XVe siècle*, Cahiers de Fanjeaux 27 (1992): 270–71.

51. Martin Aurell, "Prophétie et messianisme politique: La péninsule iberique au miroir du Liber ostensor de Jean de Roquetaillade," in *Textes prophétiques et la prophétie en Occident XIIe–XVIe siècle*, ed. André Vauchez (Rome: École Française de Rome, 1990), 43–50; John Moorman, *A History of the Franciscan Order: From Its Origins to the Year 1517* (Oxford: Clarendon Press, 1968), 49. See also Mark D. Johnston, *The Evangelical Rhetoric of Ramon Llull: Lay Learning and Piety in the Christian West Around 1300* (New York: Oxford University Press, 1996), 9–33; John August Bollweg, "Sense of a Mission: Arnau of Vilanova on the Conversion of Muslims and Jews," in *Iberia and the Mediterranean World*

3. JOHN OF RUPESCISSA'S VISION OF THE END

of the Middle Ages: Studies in Honor of Robert I. Burns, S.J., ed. Larry J. Simon (Leiden: Brill, 1995), 50–71.

52. See chapter 6, pages 125–27.

53. LSE, 117, 120. Rupescissa's discussion of the location of the ecclesiastical and secular governments is far more specific than that of Olivi. Olivi wonders whether the seat of church will move to Jerusalem, but he concludes that no answer can be derived from Scripture. See Burr, *Olivi's Peaceable Kingdom*, 192–94. Moreover, Olivi's third state is markedly more egalitarian and democratic than Rupescissa's prediction: according to Olivi, no member of the church will be subservient to another in the future age. See Lewis, "Prophet of the Year 2000," 1:241–44.

54. LSE, 125: "Vicesimo nono intellexi tantam plenitudinem Spiritus Sancti post mortem proximi Antichristi descendere super totum orbem, et maxime super Iudeos tunc conversos ad Christum et quam maxime super summos pontifices et clerum et Ecclesiam universam."

55. LSE, 122: "Vidi etiam non in eadem urbe manere continue futurum generalem Augustum *de semine Abraham* assumendum, sed extra Ierusalem pontificalem habitare debere, ne polluatur clerus sanctissimus communi convictu secularium hominum, quia non *erit* tunc *sicut populus, sic sacerdos.*"

56. LSE, 126: "et illi qui ad propagandum orbem erunt sancto matrimonio copulati vivent in maxima puritate, pro maiori parte sub tertia regula sancti Francisci, suos filios educantes in Dei timore et lege."

57. The basis for this identification in the LSE (152–58) is so striking that it merits notice here. Not only did Louis's lineage link him to Frederick II (who had been traditionally labeled an Antichrist by Joachites) but, when added together as Roman numerals, the initials of Louis's name yield 666, the number of the beast in Apoc. 13:18. See Lerner, "Historical Introduction," 56, for this helpful figure:

L	=50
V	=5
D	=500
O	=0
V	=5
I	=1
C	=100
V	=5
S	=0
	666

58. VM, prologue. "Cum enim de scripturam sacram certissime probationes summantur in confirmatione certissimorum eventuum futurorum cum declaratione et inefabili calcullatione terminorum certorum et expressorum annorum certius."

59. Reeves, *Influence of Prophecy*, ix.

3. JOHN OF RUPESCISSA'S VISION OF THE END

60. As David Burr notes in his *Persecution of Peter Olivi, Transactions of the American Philosophical Society* 66, no. 5 (Philadelphia: American Philosophical Society, 1976), 9. Edith Sylla makes the important point that theology was dependent on the development of textual analysis because the revealed word of God was accessible to humans only to the extent that humans could interpret language. The arts of the trivium were thus vital to the study of theology; the auxiliary arts eventually expanded to include naturalist study as well. See Sylla, "Autonomous and Handmaiden Science," in *The Cultural Context of Medieval Learning*, ed. John Emery Murdoch and Edith Dudley Sylla (Dordrecht: Reidel, 1975), 352.

61. Quoted in Lerner, "Ecstatic Dissent," *Speculum* 67 (1992): 33.

62. LSE, 18, 147–49; for the vision of the Virgin, see the LO, 141–43. This vision is especially interesting, considering the presence of several symbols of wisdom within it. For a useful exposition of sapiential theology and similar visions of the Virgin as the embodiment of wisdom, see Barbara Newman, *Sister of Wisdom: St. Hildegard's Theology of the Feminine* (Berkeley: University of California Press, 1987).

63. Lerner, "Ecstatic Dissent," 47. There were a few exceptions, including the twelfth-century abbess and prophet Hildegard of Bingen, who is frequently cited as a source in Rupescissa's works.

64. Lerner, "Historical Introduction," 36, citing Innocent III's *Cum ex iniuncto* of 1199.

65. LSE, 149: "Non ergo dico me prophetam missum a Deo talem sicut fuit Ysaias et Ieremias—ita ut dixerit michi Deus: 'Dic hoc populo aut illud'—sed solum dico quod Deus omnipotens aperuit michi intellectum."

66. VM, prologue: "Noscat pater firmiter amicitia vestra licet omni volenti audire denunciare terribiles eventus futuros in proximo in universo mundo me non esse prophetam missum a domino per verbum hoc dicit dominus cui enim fuerunt ysaias yer et xii sanctissimi prophete. . . . sicut enim contulit michi indignissimo intelligentiam spiritus sancti prophetalium scripturarum dominus iesu christus ad intelligenda inferius describenda."

67. Bernard McGinn, ed., *Visions of the End: Apocalyptic Traditions in the Middle Ages* (1979; New York: Columbia University Press, 1998), 4.

68. On the similarities between Rupescissa and the scholastic theologians on prophecy, see Jean-Pierre Torrell, "La conception de la prophétie chez Jean de Roquetaillade," in *Textes prophétiques et la prophétie en Occident (XIIe-XVIe siècle)*, ed. André Vauchez (Rome: École Française de Rome, 1990), 267–86.

69. Lerner, "Ecstatic Dissent," 34. Joseph Ziegler is critical of Lerner's analysis of Arnald of Vilanova and the "ecstasy defense," an explanation that Joseph Ziegler finds to be somewhat "sinister." Ziegler counters that Arnald believed that medical knowledge could derive from revelation and that he claimed divine inspiration in spiritual matters as an extension of his role as a physician; there is more on this in chapter 5. See Ziegler, *Medicine and Religion, c. 1300: The Case of Arnau of Vilanova* (Oxford: Clarendon Press, 1998), 117.

3. JOHN OF RUPESCISSA'S VISION OF THE END

70. See Bernard McGinn, *Apocalyptic Spirituality* (New York: Paulist Press, 1979), 100.

71. Reeves, *Prophetic Future*, 4.

72. Burr notes this tradition in his *Persecution of Peter Olivi*, 5, 73. Olivi is generally considered to have been more conservative than Joachim of Fiore (and, as I noted, certainly more conservative than his later disciples, including Rupescissa) in expressing any predictions that might be considered explicitly prophetic.

73. Reeves, *Influence of Prophecy*, 72.

74. McGinn, ed., *Visions of the End*, 4.

75. It is important to note that while many of Rupescissa's predictions are based on the interpretation of Scripture or other prophecies, some are straightforward pronouncements on the future without obvious textual support. Furthermore, Rupescissa had no reserve about lifting from *De mysterio cymbalorum* Arnald of Vilanova's calculation for the date of Antichrist's advent. It does not appear that Rupescissa's biblical exegesis was limited to that guided by spiritual intelligence, although he may have found it prudent to conceal this from his judges. Moreover, Rupescissa also occasionally identified himself as a prophet in the mold of Daniel or Jeremiah. For a particularly cogent example, see his remarks in the letter *Reverendissime pater*, edited by Elizabeth Casteen in "John of Rupescissa's Letter *Reverendissime pater* (1350) in the Aftermath of the Black Death," *Franciscana* 6 (2004): 158, 180–81. He makes a similar comparison of himself to the Old Testament prophets in LO, 510–37.

76. Burr, *Olivi's Peaceable Kingdom*, 118. The idea of progressive spiritual intelligence actually predates Joachim and Olivi. Robert of Liège expressed an early version of this doctrine; Isaiah 11:2–3 also suggests that spiritual intelligence would be one of the seven gifts of the Holy Spirit after the resurrection and before the end of time. See Lerner, "Ecstatic Dissent," 49. This idea of progress may also have interesting implications for a sense of advancement in scholastic learning; see Chiara Crisciani, "History, Novelty, and Progress in Scholastic Medicine," *Osiris* 2nd ser., 6 (1990): 118–39.

77. LSE, 71: "Congregabuntur ergo ad litteram *primates* ecclesiastici *et pastores* cum universa multitudine electorum ac cum strenuis militibus christianis. Consurgent enim quidam milites strenui ex semine principum gallicorum, in Machabeis fortissimis figurati, qui—quamquam prius conculcati et ad deserta fugati—in numero brevi et auxilio parvo facient validas vindictas in nationes reprobas et contra Antichristum ac si contra alterum Antiocum suscitabunt turbidas tempestates."

78. See Casteen, "John of Rupescissa's Letter," 160.

79. LSE, 65: "In diebus illis rex Francie unum ex duobus eligat: aut collum subiciat Antichristo—quod absit!—aut regno per Antichristum privetur ad tempus et patiatur patienter deiectus Antichristi temptationem cum Ecclesia sacrosancta Romana.... Honorabilius est enim regi Francie a bestia superari et cum Ecclesia pati ad tempus breve quam consentire errori. Immo rex ipse deberet tam in hiis secretis quam in antipape scandalis affuturis diligentius informari, quia minus iacula feriunt que previdentur et previsi laquei securius evitantur."

3. JOHN OF RUPESCISSA'S VISION OF THE END

80. Rupescissa cites the Tiburtine and Erythraean Sibylline prophecies, as well as Pseudo-Methodius and an *Oraculum Sibillum*, in his *Liber ostensor*. References to the *Sibylla Erythrea* and Pseudo-Methodius also appear in Rupescissa's commentary on the *Oraculum Cyrilli*. See Bignami-Odier, "Jean de Roquetaillade," 119–20. These are not the only prophetic writings that interested Rupescissa. References to the prophecies of Hildegard of Bingen also appear in both the *Liber ostensor* (in which she is cited abundantly) and the *Vade mecum* (int. 17). The unnamed prophecy of Hildegard in the *Vade mecum* is from her *Liber divinorum operum*, ed. A. Derolez and P. Dronke, Corpus Christianorum, Continuatio Medievalis, vol. 92 (Turnhout: Brepols, 1996), Visio Decima, Cap. 32.

81. VM, int. 16: "in his laboribus virilibus resistent maxime cum sciunt tempus et certas quotas annorum et tribullatio erroris nerronis et antichristi non durabit nisi duntaxat per tres annos et semis et quod consequenter est integraliter ecclesia reparanda."

82. VM, int. 16: "ita est ars certa data quod ad totidem annos est provisio fatienda in cavernis montium de fabis et leguminibus milio et carnibus salsis et fructibus sicis beatus autem qui eam faceret ad decem annos scilicet usque ad annum domini mccclxx ut sic omnes tribulationes tuti transirent."

4. ALCHEMY IN THEORY AND PRACTICE

1. John of Rupescissa, *Liber de consideratione quinte essentie omnium rerum* (Basel, 1561) (hereafter cited as DQE), 145: "Et cum ab inimicis meis iniuste in obscuro carcere tenerer, quia corpus meum a malitia ferramentorum et carceris corrumpebatur, ingeniavi, benignitate servitoris, habere aquam ardentem a quodam viro Dei et amico meo: et sola unccione cum ea facta circa locum, curatus fui."

2. *Chartularium universitatis Parisiensis*, translated in Lynn Thorndike, *University Records and Life in the Middle Ages* (New York: Columbia University Press, 1944), 246. On the influence of Aristotelian natural philosophy within the universities, see Edward Grant, *The Foundations of Modern Science in the Middle Ages: Their Religious, Institutional, and Intellectual Contexts* (Cambridge: Cambridge University Press, 1996).

3. *Statuti dell' università e dei collegii dello studio bolognese*, translated in Thorndike, *University Records*, 279–82. For an overview of the arts and medicine curricula at medieval universities, see John Baldwin, *Scholastic Culture of the Middle Ages, 1000–1300* (Lexington, Mass.: Heath, 1971); Nancy G. Siraisi, *Medieval and Early Renaissance Medicine: An Introduction to Knowledge and Practice* (Chicago: University of Chicago Press, 1990), 48–77. There was, of course, a great deal of variety in medical education. Medical knowledge did not have to be acquired within the context of the university; informal apprenticeship and private study (particularly within monasteries and houses of canons) coexisted alongside formal university training. Surgery, for instance, was a university discipline only in Italy; in other areas of Europe, surgeons operated in the medical marketplace alongside a wide variety of other non-university-trained practitioners, including barber-surgeons and empirics.

4. ALCHEMY IN THEORY AND PRACTICE

4. On the *Articella*, see Cornelius O'Boyle, *Thirteenth- and Fourteenth-Century Copies of the Ars medicine: A Checklist and Contents Description of the Manuscripts* (Cambridge: Wellcome Unit for History of Medicine, 1998). On medical education in general, see his *The Art of Medicine: Medical Teaching at the University of Paris, 1250–1400* (Leiden: Brill, 1998).

5. On alchemy's absence from university curricula, see Chiara Crisciani, "La *quaestio de alchimia* fra Duecento e Trecento," *Medioevo* 2 (1976): 119–68. Michela Pereira suggests that this exclusion may have been rooted in alchemy's dependence on practice to formulate theory, a method that ran contrary to contemporary philosophical views that practice was subordinate to theory. See her "Heaven on Earth: From the *Tabula Smaragdina* to the Alchemical Fifth Essence," *Early Science and Medicine* 5, no. 2 (2000): 131–44. See also William R. Newman, "Technology and Alchemical Debate in the Late Middle Ages," *Isis* 80 (1989): 423–45.

6. Useful surveys of alchemy include W. Ganzenmüller, *Die Alchemie im Mittelalter* (Paderborn: Verlag der Bonifaciusdruckerei, 1938), translated as *L'Alchimie au moyen âge*, trans. G. Petit-Dutaillis (Paris: Aubier, 1940); Eric John Holmyard, *Alchemy* (Harmondsworth: Penguin, 1957); Frank Sherwood Taylor, *The Alchemists: Founders of Modern Chemistry* (New York: Schuman, 1949); Robert P. Multhauf, *The Origins of Chemistry* (1966; reprint, Langhorne, Penn.: Gordon and Breach Science Publishers, 1993); Multhauf, "The Science of Matter," in *Science in the Middle Ages*, ed. David C. Lindberg (Chicago: University of Chicago Press, 1978), 369–90; Robert Halleux, *Les textes alchimiques*, Typologie des sources du moyen âge occidental 32 (Turnhout: Brepols, 1979); Michela Pereira, *L'oro dei filosofi: Saggio sulle idee di un alchimista del Trecento* (Spoleto: Centro Italiano di Studi sull'Alto Medioevo, 1992), 1–83; Stanton J. Linden, *The Alchemy Reader: From Hermes Trismegistus to Isaac Newton* (Cambridge: Cambridge University Press, 2003). Some important historiographical essays have appeared, including Barbara Obrist, "Vers une histoire de l'alchimie médiévale," *Micrologus* 3 (1995): 3–43; William R. Newman and Lawrence M. Principe, "Some Problems with the Historiography of Alchemy," in *Secrets of Nature: Astrology and Alchemy in Early Modern Europe*, ed. William R. Newman and Anthony Grafton (Cambridge, Mass.: MIT Press, 2001), 385–431; Newman and Principe, "Alchemy vs. Chemistry: The Etymological Origins of a Historiographic Mistake," *Early Science and Medicine* 3 (1998), 32–65.

7. Joseph Needham with Lu Gwei-Djen, *Spagyrical Discovery and Invention: Magisteries of Gold and Immortality*, vol. 5, pt. 2 of *Science and Civilisation in China* (Cambridge: Cambridge University Press, 1974), 114–15; Nathan Sivin, "Chinese Alchemy and the Manipulation of Time," *Isis* 67 (1976): 513–26.

8. Robert Halleux, *Papyrus de Leyde, Papyrus de Stockholm: Fragments de recettes*, vol. 1 of *Les alchimistes grecs* (Paris: Société d'Edition "Les Belles Letters," 1981).

9. For more on the contributions of Arab and Persian authors to alchemy, see Halleux, *Les textes alchimiques*, 64–69; Gerald J. Gruman, *A History of Ideas About the Prolongation of Life: The Evolution of Prolongevity Hypotheses to 1800*, Transactions of the American Philosophical Society 56:9 (Philadelphia: American Philosophical Society, 1966), 56–62. The writings of the author Jābir in Arabic are not the same as those that appeared in the Latin West

4. ALCHEMY IN THEORY AND PRACTICE

under the name of "Geber." See Pseudo-Geber, *The Summa Perfectionis of Pseudo-Geber: A Critical Edition, Translation and Study*, ed. William R. Newman (Leiden: Brill, 1991).

10. Halleux, *Les textes alchimiques*, 74–76; see also Pamela O. Long, *Openness, Secrecy, Authorship: Technical Arts and the Culture of Knowledge from Antiquity to the Renaissance* (Baltimore, Md.: Johns Hopkins University Press, 2001), 82–88.

11. Chiara Crisciani, "Alchemy and Medicine in the Middle Ages: Recent Studies and Projects for Research," *Bulletin de philosophie médiévale* 38 (1996): 19. The interpolation appears in the "old" translation of the *Meteorologica* by Alfred the Englishman, completed around 1200. By 1260, William of Moerbeke finished a new translation from the Greek, which omitted the Avicennian chapter, but *De congelatione* continued to have influence in the universities even once its true author was established.

12. Albert the Great was also critical of alchemists, whose writings he viewed as "lacking in [evidence] and proof, merely relying on authorities, and concealing their meaning in metaphorical language." See Albertus Magnus, *Book of Minerals*, trans. Dorothy Wyckoff (Oxford: Clarendon Press, 1967), esp. 167–85 (3.1. 6–9). See also Robert Halleux, "Albert le Grand et l'alchimie," *Revue des sciences philosophiques et théologiques* 66 (1982): 57–80; Pearl Kibre, "Albertus Magnus on Alchemy," in *Albertus Magnus and the Sciences, Commemorative Essays 1980*, ed. James A. Weisheipl, O.P. (Toronto: Pontifical Institutes of Mediaeval Studies, 1980), 187–202.

13. William R. Newman, "An Overview of Roger Bacon's Alchemy," in *Roger Bacon and the Sciences: Commemorative Essays*, ed. Jeremiah Hackett (Leiden: Brill, 1997), 318. Also see Gruman, *History of Ideas About the Prolongation of Life*, 58–66. On transmutatory alchemy's credibility problems in the early modern period, see Chiara Crisciani, "From the Laboratory to the Library: Alchemy According to Guglielmo Fabri," in *Natural Particulars: Nature and the Disciplines in Renaissance Europe*, ed. Anthony Grafton and Nancy Siraisi (Cambridge, Mass.: MIT Press, 1999), 295–97.

14. For a discussion of the development of the alchemical elixir, see Michela Pereira, "Teorie dell'elixir nell'alchimia latina medievale," *Micrologus* 3 (1995): 103–48; Pereira, "Un tesoro inestimabile: Elixir e 'prolongatio vitae' nell'alchimia del '300," *Micrologus* 1 (1992): 161–87.

15. John of Rupecissa, *Liber lucis*, in *Il libro della luce*, ed. Andrea Aromatico (Marsilio: Editori, 1997) (hereafter cited as LL), 19r: "Consideravi tribulationes electorum in Sacrosancto Evangelio prophetatas a Christo, maxime tribulationes pre tempore Antichristi, instare in annis in quibus est unio sub Sacrosancta Romana Ecclesia non dubium plurimum affligenda et ad montes fugienda et certe per tirannos omnibus diviciis temporalibus expolianda in brevi; sed licet iactetur fluctibus Petri navicula, liberanda tamen est in fine tribulacione dierum et generalis domina remanebit. Quapropter, ad solvendam gravem inopiam et paupertatem futuram populi sancti et electi Dei cui datum est noscere misterium veritatis, sine parabolis Lapidem Philosophorum maxime ad album et ad rubeum volo summatim dicere et impresentiarum clarissime revelabo." For this study, I use Andrea Aromatico's modern critical edition of the *Liber lucis*, based upon a fourteenth-century copy recently discovered in the private archive

4. ALCHEMY IN THEORY AND PRACTICE

of the Ubaldini family. For a description of the manuscript, see Aromatico, "Premessa al testo," in LL, 116–19.

16. Roger Bacon similarly writes in the *Opus maius* that transmutation will "contribute to the well-being of the republic, and provide everything desirable that sufficient supplies of gold can purchase." He does not, however, link transmutation to the financial needs of the persecuted church during the Antichrist's reign, as Rupescissa does. See *The "Opus Majus" of Roger Bacon*, ed. John Henry Bridges, 3 vols. (Oxford: Clarendon Press, 1897), 2:215. A connection between the state, transmutation, and Antichrist appears again in the seventeenth-century *Introitus apertus ad occlusum regis palatium* of Eirenaeus Philalethes (George Starkey) in a strikingly different form. Starkey writes that during the time of Antichrist, the philosophers' stone will be so widely used that gold will become worthless and, as a result, people will value God rather than money. This development in the eschatological role of the philosophers' stone during the era of Antichrist—Starkey's is quite an inversion of Rupescissa's position—is interesting and deserves further exploration. The passage of Starkey's *Introitus* is cited in William R. Newman, *Gehennical Fire: The Lives of George Starkey, an American Alchemist in the Scientific Revolution* (Cambridge, Mass.: Harvard University Press, 1994), 11.

17. LL, 26v: "Et aliqui, quando volunt operare ad argentum, proiciunt prius medicinam super verum argentum, et argentum sic medicinatum proiciunt supra corpus, id est supra cuprum vel ferrum, ut supra alia corpora, ad convertendum ea in verum argentum. Ad aurum vero, primo proicitur medicina supra verum aurum purissimum et est aurum medicinatum, et medicina supra cuprum et alia corpora ad convertendum ipsa corpora in verum aurum."

18. Robert P. Multhauf gives a more detailed summary of these procedures in his "John of Rupescissa and the Origin of Medical Chemistry," *Isis* 45 (1954): 361–63. He notes that this text shares its mercury formula with Pseudo-Geber. Rupescissa also gives what Multhauf characterizes as a relatively straightforward account of the process of synthesis. The *Liber lucis* attempts to identify code-named materials found in other writings with well-known substances, such as his identification of the "sulphur of the philosophers" with the well-known chemical vitriol. Rupescissa, however, is hardly the first to claim that he is the first to render clear the obfuscations of previous alchemists. See also Halleux, "Les ouvrages alchimiques de Jean de Roquetaillade," *Histoire littéraire de la France* 41 (1981): 262–67.

19. This is Multhauf's assessment in "Origin," 361–63. Rupescissa cites in addition to Hermes and Geber the alchemists Arnald of Vilanova and Alphidius. He also mentions a *Rosarius philosophorum*, although it is unclear if this is the same text that has been attributed to Arnald of Vilanova.

20. LL, 25r–v: "Credas, vir pauper evangelice, quod nullus ante me hominum ita aperte hanc veritatem scripsit, scito autem et animadverte quod multi Magistri, quibus revelatum fuit hunc misterium a revelantibus, imprecati fuerunt et maledicti horrendas maledictiones, timentes ne hec arcana venirent in manus indignorum. Ego autem has maledictiones non curo pro certo quia non revelo archanum hominibus neque filiis

4. ALCHEMY IN THEORY AND PRACTICE

hominum sceleratis, sed mistico Sacrosancto corpori Iesu Christi, videlicet Ecclesie Romane que non habet maculam mortalis criminis, nec peregrinum colorem heresie vestit, nec errorum. Hoc enim solum pro Sanctis revelavi in proxima Antichristi tempora tribulationibus remedia."

21. He also discusses the identity of the "evangelical poor," and he describes the controversy over *usus pauper* as the "scandal of the impugnment of evangelical poverty" in John of Rupescissa, *Vade mecum in tribulatione*, Venice, Biblioteca Nazionale Marciana, MS Lat. III, 39, fol. 1–19v (hereafter cited as VM), int. 14.

22. For a description of a standard secrecy clause, see Halleux, *Les textes alchimiques*, 80.

23. Multhauf, "Origin," 360. See also the more recent writings of Allen G. Debus, *The Chemical Philosophy: Paracelsian Science and Medicine in the Sixteenth and Seventeenth Centuries*, 2 vols. (New York: Science History Publications, 1977), 1:21; R. Palmer, "Pharmacy in the Republic of Venice in the Sixteenth Century," in *The Medical Renaissance of the Sixteenth Century*, ed. A. Wear, R. K. French, and I. M. Lonie (Cambridge: Cambridge University Press, 1985), 100–117; Michela Pereira, "*Medicina* in the Alchemical Writings Attributed to Raimond Lull (Fourteenth–Seventeenth Centuries)," in *Alchemy and Chemistry in the Sixteenth and Seventeenth Centuries*, ed. Piyo Rattansi and Antonio Clericuzio (Dordrecht: Kluwer Academic Publishers, 1994), 1–16. All mention Rupescissa (often along with Roger Bacon, Arnald of Vilanova, and Ramón Llull) as a precursor to Paracelsian pharmacology.

24. For the use of alchemy by surgeons in the twelfth through fourteenth centuries, see Michael R. McVaugh, "Alchemy in the *Chirurgia* of Teodorico Borgognoni," in *Alchimia e medicina nel medioevo*, ed. Chiara Crisciani and Agostino Paravicini Bagliani (Florence: Sismel, 2003), 55–75.

25. See chapter 1, page 3. John of Rupescissa, *Johannes' de Rupescissa "Liber de consideratione quintae essentiae omnium rerum" deutsch: Studien zur Alchemia medica des 15. bis 17. Jahrhunderts mit kritischer Edition des Textes*, ed. Udo Benzenhöfer (Stuttgart: Franz Steiner, 1989), 55–82. See also Multhauf, "Origin," 366; "Science of Matter," 379.

26. Multhauf, "Origin," 366.

27. DQE, 11: "Consideravi ergo quod tempus expensum in pruritu mundanae Philosophiae plusquam annis quinque ante ingressum ordinis in florentissimo studio Tolosano, et plus aliis quinque ex quo ordinem hunc intravi cum strepitu multo inanium verborum, et conflictu inutilium disputationum, et laude vana et gloria fatua lectionum, tam in studiis particularibus, quam in generalibus legens."

28. DQE, 12: "Et vidi possibilem modum in hoc, si utilitates quas vidi in Philosophia divino spiritu illustrante, etiam tempore iuventutis meae, quo tempore assuit mihi Deus misericors et miserator supra modum, ut revelem pauperibus Christi et Evangelicis viris, ut qui divitias propter Evangelium contempserunt, sine humana doctrina sciant faciliter, et de levi, ac sine notabilibus expensis, suas corporales miserias et infirmitates humanas, divina benignitate sanare: impedimenta sanctae orationi et meditationi mirabiliter effugare, tentationibus etiam daemonum, quas propter quasdam infirmitates

4. ALCHEMY IN THEORY AND PRACTICE

inferunt, resistere: ut in omnibus et per omnia expedite possint totis viribus Domino nostro Jesu Christo devotissime servire. Et cum istud opus pro solis Sanctis faciam, a quo opere, imperante Deo, recipient tanta bona, concluditur evidenter quod ero in eorum beneficiis, orationibus, et meritis quae amplius facient auxilio libro huius."

29. DQE, 13: "Et sic ex tempore expenso in Philosophia cum multa vanitate, mihi non cessabunt nova merita pervenire, non solum decem annis in Philosophia pertransitis, sed plusquam mille annis ante finem seculi sine dubio defluxuris."

30. Isaac Israeli, *Omnia opera Ysaac*, 2 vols. (Leyden, 1515). *De virtutibus simplicium medicinarum* is now usually attributed to Constantine the African or the Salernitan doctor John of Saint Paul. On the latter, see V. L. MacKinney, "*Dynamidia* in Medieval Medical Alchemy," in *Isis* 24 (1936): 412. See also Halleux, "Ouvrages," 256. While such reproduction of authoritative texts was common, it points to Rupescissa's general preference to focus his efforts on theoretical elaboration rather than on personal observations of therapeutic effect, as I discuss below.

31. DQE, 127: "Vel quando pauperes Evangelici vadunt praedicando per mundum, et laeduntur in pedibus, cum solo tartaro calcinato statim sanabuntur: quia habet vim quintae Essentiae, cum sit eiusdem naturae. Nam aqua divinae actionis, omnem impetiginem et serpiginem sanat."

32. DQE, 55: "ad servandum in toto tempore bellorum et tribulationum, et maxime tempore Antichristi."

33. DQE, 134: "ideo est cautela ut princeps populi Christiani in ordine bellorum habeat sic in doliis aquam ardentem paratam, ut cuilibet pugili tribuat medium scyphum vel circa in principio bellicosi congressus: et debet hoc arcanum omnibus inimicis Ecclesie occultari: imo nec princeps, nec alii ministrantes debent hoc alicui revelare."

34. DQE, 15: "Quoniam paucissimi Philosophi pervenerunt ad ultimam notitiam rei talis, evidens est, et per hoc quod medici nostri temporis, qui ardent desiderio pecuniae et honorum, pro pecunia nil tale alicui magnati nostri temporis potuerunt conferrre nec possunt: nec vult Deus ut avari sciant." But as Chiara Crisciani and Michela Pereira point out, some contemporary physicians (such as Gentile da Foligno and Maino de' Maineri) were aware of the distilled remedies that Rupescissa described in *De quinta essentia*. See "Black Death and Golden Remedies: Some Remarks on Alchemy and the Plague," in *The Regulation of Evil: Social and Cultural Attitudes to Epidemics in the Late Middle Ages*, ed. Agostino Paravicini Bagliani and Francesco Santi (Florence: Sismel, 1998), 20.

35. His maladies and their alchemical cure are described DQE, 144–45, and Jean de Roquetaillade (John of Rupescissa), *Liber ostensor quod adesse festinant tempora*, ed. Clémence Thévenaz Modestin, Christine Morerod-Fattebert, André Vauchez, et al. (Rome: École Française de Rome, 2005), 515–26. For the seeming incompatibility of empirical naturalism and Joachite inspiration, see Paola Zambelli, introduction to *'Astrologi hallucinati:' Stars and the End of the World in Luther's Time*, ed. Paola Zambelli (Berlin: Walter de Gruyter, 1986), 20.

36. On the relationship between medical theory and practice, see Nancy G. Siraisi's chapter on disease and treatment in her *Medieval and Early Renaissance Medicine*, 115–52. See

also E. Ruth Harvey, *The Inward Wits: Psychological Theory in the Middle Ages and the Renaissance* (London: Warburg Institute, 1975), for practicing medieval physicians' tendency to rely upon Galenic and Stoic interpretations of the location of the intellect, although these interpretations often clashed with those of contemporary scholastic philosophers. John Buridan also recommended that readers rely on information derived from the senses for use in medicine rather than on general principles derived deductively. See Jack Zupko, *John Buridan: Portrait of a Fourteenth-Century Arts Master* (Notre Dame, Ind.: University of Notre Dame Press, 2003), 203–26.

37. Michael R. McVaugh, "Moments of Inflection: The Careers of Arnau de Vilanova," in *Religion and Medicine in the Middle Ages*, ed. Peter Biller and Joseph Ziegler (York: York Medieval Press, 2001), 49–50.

38. For Aquinas and William of Conches, see M. D. Chenu, "Nature and Man: The Renaissance of the Twelfth Century," in *Nature, Man, and Society in the Twelfth Century*, trans. Jerome Taylor and Lester K. Little (Chicago: University of Chicago Press, 1968), 11–12. Rupescissa refers to John Mesüe's *De simplicibus*, part of the *De medicamentorum purgantium delectu, castigatione et usu*. See Halleux, "Ouvrages," 253.

39. DQE, 25: "Et illi qui in veritate intelligunt per incomprehensibile lumen Dei causam rerum mirandarum, quas mundani Philosophi nescierunt, a sequacibus eorum inanes et phantastici reputantur." Rupescissa here alludes to accusations that he was insane or heretical, both charges leveled at him during his trial before Pope Clement VI in 1349, who found him to be be *fantasticus* rather than *hereticus*. Rupescissa also complained in *Liber ostensor* of being mocked as *fantasticus* at a public hearing in 1354. See Robert E. Lerner, "Historical Introduction," in *Liber secretorum eventuum* by John of Rupecissa, ed. Robert E. Lerner and Christine Morerod-Fattebert (Fribourg: Editions Universitaires, 1994), 30–31n. 43.

40. See chapter 7, pages 145–46, for the opinions of Thomas Aquinas and Giles of Rome.

41. DQE, 164–65: "Est autem sciendum quod homines tripliciter mori possunt: uno modo in termino constituto nobis a Deo morte naturali quem aliquo ingenio naturali praeterire non possumus: alio modo morte violenta: et istis duobus modis medicina est frustra: alio modo casualiter, vel occasionaliter, citra terminum a Deo nobis constitutum: ut qui nimia repletione vel dissolutione, vel austera abstinentia, vel desperatione, vel negligentia vitandi mortis periculum, se occidunt."

42. Luis García-Ballester, "On the Origin of the 'Six Non-natural Things' in Galen," in *Galen and Galenism: Theory and Practice from Antiquity to the European Renaissance*, ed. Jon Arrizabalaga et al. (Aldershot: Variorum, 2002), 4:105–15. See also Siraisi, *Medieval and Early Renaissance Medicine*, 120–23.

43. DQE, 14: "Secretum primum est, quod per virtutem quam Deus contulit naturae conditae, et humano magisterio subiectae, potest homo incommoda senectutis quibus nimis in antiquis viris Evangelicis impediuntur opera Evangelicae vitae, curare, et amissam iuventutem iterum restaurare, et vires pristinas recuperare, et rehabere.... Hoc est in quo la[bo]raverunt omnes quaerere rem creatam ad hominum usum aptam, quae

4. ALCHEMY IN THEORY AND PRACTICE

possit corruptibile corpus a putrefactione servare: servatum, sine diminutione conservare: conservatum, si foret possibile, perpetuare in esse: quia hoc est quod naturaliter omnes desiderant nunquam corrumpi, nec mori."

44. DQE, 17–18: "Ergo radix vitae, est quaerere rem de se (si staret in aeternum) incorruptibilem: quae omnem rem sibi unitam, et maxime carnem, semper teneat incorruptam: quae virtutem vitae ac spiritum nutriat, et augeat, et restauret: quae omne crudum digerat, et omne digestum ad aequalitatem reducat: et omnem excessum cuiuscunque qualitatis amputet, et quamcunque qualitatem deperditam restauret: humidum naturale faciat abundare, et ignem naturalem debilem inflammare procuret."

45. DQE, 19–20: "oportet rem quaerere, quae sic se habeat respectu quatuor qualitatum, quibus compositum est corpus nostrum, sicut se habet caelum respectu quatuor elementorum: Philosophi autem vocaverunt caelum quintam Essentiam, respectu quatuor elementorum, quia in se caelum, est incorruptibile et immutabile et non recipiens peregrenas impressiones, nisi fieret iussu Dei. Sic et res quam quaerimus est respectu quatuor qualitatum corporis nostri: quinta Essentia, in se incorruptibilis sic facta, non calida sicca cum igne, nec humida frigida cum aqua, nec calida humida cum aere, nec frigida sicca cum terra: sed est Essentia quinta, valens ad contraria, sicut caelum incorruptibile: quod, quando necesse est, influit qualitatem humidam, aliquando calidam aliquando frigidam, aliquando siccam."

46. Gruman, *History of Ideas About the Prolongation of Life*, 16; Multhauf, "Science of Matter," 370. For Aristotle on cosmic permanence and the fifth body, see *De caelo*, 1.1.269a30–270b32; *Meteorologica*, 1.339a11–339b26; and the Pseudo-Aristotelian *De mundo*, 392a5–b4. All citations to Aristotle can be found in *The Complete Works of Aristotle*, ed. Jonathan Barnes (Princeton, N.J.: Princeton University Press, 1984).

47. See Frank Sherwood Taylor, "The Idea of the Quintessence," in *Science, Medicine, and History: Essays on the Evolution of Scientific Thought and Medical Practice written in Honour of Charles Singer*, ed. E. Ashworth Underwood (London: Oxford University Press, 1953), 247–56.

48. For a guide to the wide range of medical practitioners and their use of the learned tradition of medicine, see Michael R. McVaugh, *Medicine Before the Plague: Practitioners and Their Patients in the Crown of Aragon, 1285–1345* (Cambridge: Cambridge University Press, 1993), 38–40; Katharine Park, *Doctors and Medicine in Early Renaissance Florence* (Princeton, N.J.: Princeton University Press, 1985), 8; Danielle Jacquart, "Medical Practice in Paris in the First Half of the Fourteenth Century," in *Practical Medicine from Salerno to the Black Death*, ed. Luis García-Ballester et al. (Cambridge: Cambridge University Press, 1994), 186–210; Charles Webster and Margaret Pelling, "Medical Practitioners," in *Health, Medicine, and Mortality in the Sixteenth Century* (Cambridge: Cambridge University Press, 1979), 165–235. For the various forms of training and licensing for practitioners, see Siraisi, *Medieval and Early Renaissance Medicine*, 17–36. On the competing interests of physicians and surgeons in fourteenth-century France, see Marie-Christine Pouchelle, *The Body and Surgery in the Middle Ages*, trans. Rosemary Morris (Oxford: Polity Press, 1990), 13–22.

49. Siraisi, *Medieval and Early Renaissance Medicine*, 71. See also Oswei Temkin, *Galenism: Rise and Decline of a Medical Philosophy*. (Ithaca: Cornell University Press, 1973).

4. ALCHEMY IN THEORY AND PRACTICE

50. Per-Gunnar Ottosson, *Scholastic Medicine and Philosophy: A Study of Commentaries on Galen's Tegni (ca. 1300–1450)* (Naples: Bibliopolis, 1984), 135. On medical grades, see Michael R. McVaugh's introduction to Arnald of Vilanova, *Aphorismi de gradibus*, ed. Michael R. McVaugh et al. (Granada: Seminarium Historiae Medicae Granatensis, 1975), 3–136.

51. For instance, DQE, 61–62: "O quam magna scientia est ista, ut scias naturam rerum comestibilium: ut intelligas quid debes comedere, quid vitare: quoniam si nimis es frigidus, utaris istis que sunt calida in primo gradu. . . . Si propter infirmitatem provenientem ex nimia frigiditate, non sufficit tibi caelum tuum in praecedenti canone factum calidum in primo gradu, extrahe quintam Essentiam ex his rebus quae sunt inferius, quae sunt calidae in secundo gradu."

52. DQE, 75: "Cum ergo scire volueris de aliqua re ex praedictis, utrum sit calida et humida, sicca et calida, frigida et humida, frigida et sicca: quaere eam in tabulis calidorum et si ibi invenies eam, extrahe vel nota eam; deinde eandem quaere in tabulis humidorum, et si ibi eam invenies, extrahe; et ita invenies complexionem et gradum eius: exempli gratia quaeras argentum vivum in tabulis calidorum, et invenies eum calidum in quarto gradu, quaeras idem in tabulis humidorum, et invenies in quarto gradu, id est in summo calidum et humidum." Rupescissa's instruction to look up substances in tables according to quality may be a reference to the *Tacuinum sanitatis* of Ibn Botlân, which includes such tables. A version has been edited by Luisa Cogliati Arano as *The Medieval Health Handbook: Tacuinum sanitatis* (New York: George Braziller, 1976).

53. For instance, DQE, 78–80. For a nuanced discussion of the differences between humors, qualities, and elements, see Ottosson, *Scholastic Medicine*, 129–66.

54. DQE, 131.

55. On radical humidity, see Thomas Hall, "Life, Death, and the Radical Moisture," *Clio medica* 6 (1971): 3–23; Michael R. McVaugh, "The *humidum radicale* in Thirteenth-Century Medicine," *Traditio* 30 (1974): 259–83.

56. Gruman, *History of Ideas About the Prolongation of Life*, 17. Some sections of the *Canon* also circulated outside of a university context, and the text includes long sections on remedies. See Nancy G. Siraisi, "Avicenna and the Teaching of Practical Medicine," in *Medicine and the Italian Universities*, ed. Nancy G. Siraisi (Leiden: Brill, 2001), 63–78.

57. See for instance, James J. Bono, *The Word of God and the Languages of Man: Interpreting Nature in Early Modern Science and Medicine*, vol. 1: *Ficino to Descartes* (Madison: University of Wisconsin Press, 1995), 29–47, 97–103.

58. Taddeo Alderotti, for instance, devoted seven *consilia* to the healing powers of alcohol distilled from wine, a substance that he described in highly effusive terms. See Nancy G. Siraisi, *Taddeo Alderotti and His Pupils* (Princeton, N.J.: Princeton University Press, 1981), 301. For an overview of the history of alcohol and healing waters, see Lu Gwei-Djen, Joseph Needham, and Dorothy Needham, "The Coming of Ardent Water," *Isis* 19 (1972): 69–112.

59. DQE, 58–59: "Deus caeli talem virtutem contulit quintae Essentiae, ut extrahat ab omni fructu, et ligno, et radice, et flore, et herba, et carne, et semine, et a qualibet rerum specie, et ex quacunque medicinali re, omnes virtutes et proprietates et naturas,

et effectus, quos Deus gloriae, conditor naturae, creavit in eis. . . . et melior in centuplo ratione quintae Essentiae, quam esset sine ea."

60. Multhauf, "Origin," 366; Halleux, "Ouvrages," 274–76.

61. Crisciani, "Alchemy and Medicine," 13.

62. Rupescissa may have been inspired to use organic products as alchemical bases by reading Jābirian texts, which he may have encountered directly or through the conduit of other works, such as the writings of Roger Bacon, the Avicennian *De anima in arte alchimiae*, the *Flos florum* or *Testamentum* attributed to Arnald of Vilanova, or the *Testamentum* attributed to Ramon Llull.

63. See (Pseudo-)Arnald of Vilanova, *Le Rosier alchimique de Montpellier (Lo Rosari): Textes, traduction, notes, et commentaires*, ed. Antoine Calvet (Paris: Presses de l'Université de Paris-Sorbonne, 1997).

64. Astrology and astronomy were closely related in the Middle Ages, inasmuch as belief in the influence of the heavens was widespread and competence in astrology was often the goal of astronomical study. Basic university instruction in astronomy via *On the Sphere* by John of Sacrobosco or the *Theorica planetarum* did not involve astrology, however. Nancy Siraisi notes that although astrology was not an official part of the medical curriculum, introductory astronomy and astrology were important elements in the training of Bolognese medical students in the early fourteenth century; see her *Taddeo Alderotti*, 139–46. H. Darrel Rutkin has recently argued that astrology was taught formally within medical curricula at Bologna, Padua, Ferrara, and Paris, among other universities; see his "Astrology, Natural Philosophy, and the History of Science, c. 1250–1700: Studies Toward an Interpretation of Giovanni Pico della Mirandola's *Disputationes adversus astrologiam divinatricem*," Ph.D. diss., Indiana University, 2002, esp. chapter 3. Danielle Jacquart shows that the implementation of astrology by physicians was variable: some used it extensively, some not at all. See her "Medical Scholasticism," in *Western Medical Thought from Antiquity to the Middle Ages*, ed. Mirko D. Grmek (Cambridge, Mass.: Harvard University Press, 1998), 233–35.

65. Roger French, "Astrology in Medical Practice," in *Practical Medicine from Salerno to the Black Death*, ed. Luis García-Ballester et al. (Cambridge: Cambridge University Press, 1994), 30–59. See for instance, Alberto Alonso Guardo, *Los pronósticos médicos en la medicina medieval: El "Tractatus de crisi et de diebus creticis" de Bernardo de Gordonio*, Lingüística y Filología 54 (Valladolid: Secretariado de Publicaciones e Intercambio Editorial, Universidad de Valladolid, 2003). Also see Cornelius O'Boyle, *Medieval Prognosis and Astrology: A Working Edition of Aggregationes de crisi et creticis diebus, With Introduction and English Summary* (Cambridge: Wellcome Unit for the History of Medicine, 1991).

66. McVaugh, *Medicine Before the Plague*, 164–65.

67. On Arnald's use of astrological seals and talismans, see Nicolas Weill-Parot, "Arnaud de Villeneuve et les relations possibles entre le sceau du Lion et l'alchimie," *Arxiu de textos catalans antics* 23/24 (2004–2005): 269–80.

68. An excerpt from this treatise is published in Rosemary Horrox, ed., *The Black Death* (Manchester: Manchester University Press, 1994), 158–63. Also see Jon Arrizabalaga,

"Facing the Black Death: Perceptions and Reactions of University Medical Practitioners," in *Practical Medicine from Salerno to the Black Death*, ed. Luis García-Ballester et al. (Cambridge: Cambridge University Press, 1994), 241–53. Danielle Jacquart, however, argues that the statement of the faculty of medicine, the *Compendium de epidemia*, mentions the planetary conjunction only once and focuses more closely on the corruption of the air. See her "Medical Practice in Paris," 204–5.

69. William R. Newman and Anthony Grafton, introduction to *Secrets of Nature: Astrology and Alchemy in Early Modern Europe*, ed. William R. Newman and Anthony Grafton (Cambridge, Mass.: MIT Press, 2001), 15–19.

70. See my discussion of *Decknamen* in chapter 6 below.

71. DQE, 42: "quod natura illam quintam Essentiam sic perficit et perfecit, ut sine alia praeparatione ex venis transferat ipsum sanguinem immediate in carnem: et hanc quintam Essentiam tam propriam naturae maximum est haberi: quia in ea est virtus mirabilis caeli stellati nostri: et ad curam naturae divinissima operatur miracula, sicut infra docebo."

72. DQE, 32–33: "cum foramen aperueris, si odor, qui super mirabilis esse debet, ita quod nulla mundana fragrantia ei valeat adaequari, qui videbitur descendisse quasi de sublimitate gloriosissimi Dei, intantum quod si fuerit vas positum in angulo domus, ex fragrantia quintae Essentiae (quod mirabile et summe miraculosum est) attrahat ad se vinculo invisibili universos intrantes." (Emphasis is mine.)

73. DQE, 104–5: "deinde videbis miraculum stupendum, quoniam per rostrum alembici videbis quasi mille venulas liquorum benedictae minerae descendere per guttas rubeas sicut recte esset sanguis.... Si dicerem tibi millies, hoc est secretum secretorum, nunquam posset sufficere ad exprimendum medietatem arcani: aufert dolorem omnium vulnerum et mirabiliter sanat. Virtus eius est incorruptibilis et miraculosa et utilis supra modum." (Emphasis is mine.) Rupescissa uses the terms "*miraculum*" or "*miraculose*" or "*quasi miraculose*" frequently to describe the quintessence's healing effects; see, for instance, DQE, 43, 82, 116, 117, 120, 123, 152, 154, 162, etc. For general definitions of "*mirabilia*" and "*miracula*," see Caroline Walker Bynum, "Wonder," in *Metamorphosis and Identity* (New York: Zone Books, 2001), 41–50; Benedicta Ward, *Miracles and the Medieval Mind: Theory, Record, and Event, 1000–1215* (Philadelphia: University of Pennsylvania Press, 1982), 3–19.

74. DQE, 120: "docebo in hoc secundo libro sanitatis remedia subito procurare et quasi miraculose Evangelicis viris."

75. See Crisciani and Pereira, "Black Death and Golden Remedies," 20–21.

76. DQE, 41–42: "Si audivisti quod tibi superius ex secretis maximis revelavi, cum dixi, quod in omni re quinta Essentia est, et remanet incorrupta, maximum erit tibi si docuero ipsam extrahere a sanguine humano, et a carnibus omnium animalium, et ab ovis, et a similibus rebus."

77. Pereira, "Heaven on Earth"; Tullio Gregory, *Anima mundi: La filosofia di Guglielmo di Conches e la Scuola di Chartres* (Florence: G. C. Sansoni, 1955), 95; M. Lapidge, "The Stoic

4. ALCHEMY IN THEORY AND PRACTICE

Inheritance," in *A History of Twelfth-Century Western Philosophy*, ed. Peter Dronke (Cambridge: Cambridge University Press, 1988), 103.

78. Taylor, "Idea of the Quintessence, 249–51.

79. Pereira, "Heaven on Earth," 139–40.

80. DQE, 22–23: "Et sicut caelum summum non influit solum per se conservationem in mundo et influentias miras, sed per virtutem solis et aliarum stellarum: sic et caelum istud, quinta Essentia vult ornari sole mirabili, splendido et incorruptibili aequato: in quem solem non possit etiam agere ipse ignis, ut ipsum corrumpat. Et dico tibi in charitate nonficta et conscientia bona, quod iste sol illuminatus, splendidus, et incorruptibilis ab igne, qui influit incorruptibilitatem et radicem vitae eo modo quo est possibile, ut supra explicavi in corpore nostro, qui est creatus ad ornatum caeli nostri, et ad augendum influentiam quintae Essentiae, potest capi manu, et constituit eum Rex gloriae in potestate mortalium: Et ego propter charitatem Dei qui loquor Evangelicis viris, eum tibi nomine proprio et intelligibili revelabo, et ipsum est aurum Dei, quod ex lapide Philosophorum componitur.... Sol quippe est filius solis caeli, ex quo componitur lapis Philosophorum. Generatur enim ab influxibus solis in visceribus terrae, et sibi sol ex influentia sua tribuit naturam et colorem et incorruptibilem substantiam."

81. Aristotle, *De generatione et corruptione*, 2.10.336a15–30.

82. Agostino Paravicini Bagliani, "Rajeunir au moyen âge: Roger Bacon et le mythe de la prolongation de la vie," *Revue médicale de la Suisse Romande* 106 (1986): 17; Crisciani and Pereira, "Black Death and Golden Remedies," 39. A direct relationship between gold and the sun also appears in the *Tractatus* of "Solemnis Medicus." See Crisciani and Pereira, "Black Death and Golden Remedies," 33–34.

83. See, for instance, the prescription of potable gold in the *De retardatione accidentium senectutis*, an influential prolongevity text formerly attributed to Roger Bacon but actually the work of an anonymous thirteenth-century author attached to the papal court, in Roger Bacon, *Fratris Rogeri Bacon De retardatione accidentium senectutis cum aliis opusculis de rebus medicinalibus*, ed. A. G. Little and E. Withington (Oxford: Clarendon Press, 1928). See also the writings of the physician Bernard of Gordon, who recommends ambergris, potable gold, and flowers of bugloss in his *Tractatus de marasmode*: Luke Demaitre, "The Care and Extension of Old Age in Medieval Medicine," in *Aging and the Aged in Medieval Europe*, ed. Michael M. Sheehan (Toronto: Pontifical Institute of Mediaeval Studies, 1990), 19–20. See also his *Doctor Bernard de Gordon: Professor and Practitioner* (Toronto: Pontifical Institute of Mediaeval Studies, 1980)

84. On the distinction between natural gold, "alchemical gold"—which is, according to Rupescissa, corrosive—and the "gold of God" made from the philosophers' stone, see my discussion in chapter 7.

85. Crisciani and Pereira, "Black Death and Golden Remedies," 7–8.

86. Nancy Siraisi offers a different perspective, emphasizing institutional continuity and noting that patients never lost any long-term faith in the medical profession. See Park, *Doctors and Medicine*, 34–42; Nancy G. Siraisi, "The Physician's Task: Medical

4. ALCHEMY IN THEORY AND PRACTICE

Reputations in Humanist Collective Biographies," in *The Rational Arts of Living*, ed. A. C. Crombie and Nancy G. Siraisi (Northhampton, Mass.: Smith College Studies in History, 1987), 108–9; Siraisi, *Medieval and Early Renaissance Medicine*, 42–43; Crisciani, "Alchemy and Medicine," 18; Crisciani and Pereira, "Black Death and Golden Remedies," 7–39.

87. Crisciani and Pereira, "Black Death and Golden Remedies," 11–12. See also the many remedies collected by Rosemary Horrox in *The Black Death*. John of Burgundy, for instance, recommends consuming vinegar, wine, pomegranates, quinces, sandalwood, cucumbers, fennel, ginger, cinnamon, saffron, and cumin, among other foods.

88. Theriac was reputedly made from viper's flesh and other ingredients (and a version of it appears in *De retardatione*, which recommends viper's flesh as a life-prolonging medicine). It was thought to be a universal antidote to poison and a possible cure for a range of ailments. It is the subject of a treatise by Arnald of Vilanova, *Epistola de dosi tyriacalium medicinarum*, ed. Michael R. McVaugh, in vol. 3 of *Arnaldi de Villanova Opera medica omnia* (Barcelona: Seminarium Historiae Medicae Cantabricensis, 1985), 55–91 (including an introduction in English).

89. DQE, 23.

5. ARTISTS AND THE ART

1. Wilfrid Theisen, O.S.B., "The Attraction of Alchemy for Monks and Friars in the Thirteenth–Fourteenth Centuries," *American Benedictine Review* 46 (1995): 239–53.

2. Agostino Paravicini Bagliani, *The Pope's Body*, trans. David S. Peterson (Chicago: University of Chicago Press, 2000); Bagliani, "Ruggero Bacone e l'alchimia di lunga vita: Riflessioni sui testi," in *Alchimia e medicina nel medioevo*, ed. Chiara Crisciani and Agostino Paravicini Bagliani (Florence: Sismel, 2003), 33–54. See also Chiara Crisciani, *Il Papa e l'alchimia: Felice V, Guglielmo Fabri e l'elixir* (Rome: Viella, 2002).

3. Robert Halleux, "Les ouvrages alchimiques de Jean de Roquetaillade," *Histoire littéraire de la France* 41 (1981): 250; Paravicini Bagliani, *The Pope's Body*, 227.

4. On Roger Bacon in general, see Jeremiah Hackett, ed. *Roger Bacon and the Sciences: Commemorative Essays* (Leiden: Brill, 1997); Stewart C. Easton, *Roger Bacon and His Search for a Universal Science* (New York: Columbia University Press, 1952); David C. Lindberg, "Science as Handmaiden: Roger Bacon and the Patristic Tradition," *Isis* 78 (1987): 518–36; Paul L. Sidelko, "The Condemnation of Roger Bacon," *Journal of Medieval History* 22 (1996): 69–81. For Bacon's alchemy, see Dorothea Waley Singer, "Alchemical Writings Attributed to Roger Bacon," *Speculum* 7 (1932): 80–86; E. Brehm, "Roger Bacon's Place in the History of Alchemy," *Ambix* 23 (1976): 53–58; William R. Newman, "The Alchemy of Roger Bacon and the *Tres Epistolae* Attributed to Him," in *Comprendre et maîtriser la nature au moyen âge: Melanges d'histoire des sciences offerts à Guy Beaujouan*, Hautes études médiévales et modernes 73 (Geneva: Droz, 1994), 461–79; William R. Newman, "The Philosopher's Egg: Theory and Practice in the Alchemy of Roger Bacon," *Micrologus* 3

5. ARTISTS AND THE ART

(1995): 75–101. For Bacon's writings on medicine, see Faye Marie Getz, "The Faculty of Medicine Before 1500," in *Late Medieval Oxford*, ed. J. I. Catto and Ralph Evans, vol. 2 of the *History of the University of Oxford* (New York: Clarendon Press, 1992), 373–406; Getz, "To Prolong Life and Promote Health: Baconian Alchemy and Pharmacy in the English Learned Tradition," in *Health, Disease, and Healing in Medieval Culture*, ed. S. Campbell (New York: St. Martin's Press, 1992), 135–45. For Bacon's relationship to the *Secretum secretorum*, see Steven J. Williams, *The Secret of Secrets: The Scholarly Career of a Pseudo-Aristotelian Text in the Latin Middle Ages* (Ann Arbor: University of Michigan Press, 2003).

5. For Roger Bacon and the prolongation of life, see Agostino Paravicini Bagliani, "Rajeunir au moyen âge: Roger Bacon et le mythe de la prolongation de la vie," *Revue médicale de la Suisse Romande* 106 (1986): 9–23; Paravicini Bagliani, "Storia della scienza e storia della mentalità: Ruggero Bacone, Bonifacio VIII e la teoria della 'prolongatio vitae,'" in *Aspetti della letteratura latina nel secolo XIII*, ed. C. Leonardi and G. Orlandi (Florence: La Nuova Italia, 1986), 243–80.

6. Paravicini Bagliani, "Storia della scienza," 256–57.

7. Steven J. Williams, "Roger Bacon and His Edition of the Pseudo-Aristotelian *Secretum secretorum*," *Speculum* 69 (1994): 57–73; Singer, "Alchemical Writings," 80–86.

8. Roger Bacon, *Opus tertium*, vol. 1 of *Opera quaedam hactenus inedita*, ed. J. S. Brewer (London: Longman, Green, Longman and Roberts, 1859), 210. The idea that human life had shortened since remote antiquity because of neglect of health and morals is ancient. Peter of Abano (d. 1303) devoted a section of his *Conciliator* to the same subject. He cites Galen's commentary on the Aphorisms of Hippocrates, which makes a similar point. My thanks to Nancy Siraisi for bringing this to my attention.

9. Roger Bacon, *The "Opus Majus" of Roger Bacon*, ed. John Henry Bridges, 3 vols. (Oxford: Clarendon Press, 1897) (hereafter cited as OM), 2:205–7: "sed propter has duas causas abbreviata est longaevitas hominis contra naturam. Praeterea certis experimentis probatum est, quod ista festinatio nimia est retardata pluries, et longaevitas prolongata per multos annos per experientias secretas; et multi hoc scribunt auctores. Propter quod oportet quod sit haec nimia festinatio accidentalis, habens remedium possibile.... Regimen ergo sanitatis sufficiens, quantum homo possit habere, prolongaret vitam ultra communem terminum vivendi accidentalem, quem homo propter stultitiam suam sibi non servat; et sic vixerunt aliqui per multos annos ultra communem statum vivendi. Sed regimen speciale per remedia retardantia dictum communem statum quem ars regendi sanitatem non transgreditur, potest longe plus vitam prolongare."

10. OM, 2:208.

11. These remedies are to be found in *De retardatione accidentium senectutis*, a prolongevity tract attributed falsely to Bacon and likely a source for his work and available as Roger Bacon, *Fratris Rogeri Bacon De retardatione accidentium senectutis cum aliis opusculis de rebus medicinalibus*, ed. A. G. Little and E. Withington (Oxford: Clarendon Press, 1928). On Bacon's relationship to the treatise, see Paravicini Bagliani, *The Pope's Body*, 200–205.

12. OM, 2:215: "Nam illa medicina, quae tolleret omnes immunditias et corruptiones metalli vilioris, ut fieret argentum et aurum purissimum, aestimatur a

sapientibus posse tollere corruptiones corporis humani in tantum, ut vitam per multa secula prolongaret."

13. Roger Bacon, *Secretum secretorum cum glossis et notulis*, vol. 5 of *Opera hactenus inedita Rogeri Baconi*, ed. Robert Steele (Oxford: Clarendon Press, 1909), 114–15. For an analysis of this passage, see William R. Newman, "An Overview of Roger Bacon's Alchemy," in *Roger Bacon and the Sciences: Commemorative Essays*, ed. Jeremiah Hackett (Leiden: Brill, 1997), 328–32.

14. Paravicini Bagliani, "Rajeunir au moyen âge," 14–16.

15. See Gerald J. Gruman, *A History of Ideas About the Prolongation of Life: The Evolution of Prolongevity Hypotheses to 1800*, Transactions of the American Philosophical Society 56:9 (Philadelphia: American Philosophical Society, 1966), 60–62; Paravicini Bagliani, "Rajeunir au moyen âge," 16–18.

16. Rupescissa cites a *Rosarius* in his *Liber lucis*, fol. 28v, but he does not cite Bacon by name. For the presence of Baconian alchemy in both Arnald's and the fourteenth-century English alchemist John Dastin's versions of the *Rosarius*, see Michela Pereira, "*Medicina* in the Alchemical Writings Attributed to Raimond Lull (Fourteenth–Seventeenth Centuries)," in *Alchemy and Chemistry in the Sixteenth and Seventeenth Centuries*, ed. Piyo Rattansi and Antonio Clericuzio (Dordrecht: Kluwer Academic Publishers, 1994), 5–6.

17. Paravicini Bagliani, *The Pope's Body*, 205.

18. Roger Bacon, *Opus minus*, vol. 2 of *Opera quaedam hactenus inedita Rogeri Baconi*, ed. J. S. Brewer (London: Longman, Green, Longman, and Roberts, 1859), 373. On holy women and food, see Caroline Walker Bynum, *Holy Feast and Holy Fast: The Religious Significance of Food to Medieval Women* (Berkeley: University of California Press, 1987), 196, 274; see also Bynum, *The Resurrection of the Body in Western Christianity, 200–1336* (New York: Columbia University Press, 1995), 326–27.

19. John of Rupescissa, *Liber de consideratione quinte essentie omnium rerum* (Basel, 1561) (hereafter cited as DQE), 120.

20. Although Bacon entitles a chapter of his *Opus tertium* the "Science of the Quintessence," the chapter actually discusses demonic magic, astrology, and explosives, rather than a distilled fifth element. Bacon's chapter juxtaposes the problem of Antichrist with the phrase "*quinte essentie*," however; this coupling may have been suggestive to Rupescissa if he indeed read the work.

21. Roger Bacon, *In libro sex scientarum in 3° gradu sapientie*, in *De retardatione accidentium senectutis cum aliis opusculis de rebus medicinalibus*, ed. A.G. Little and E. Withington (Oxford: Clarendon Press, 1928), 184–85. See also Newman, "Philosopher's Egg," 78–79.

22. OM, 2:212: "Nam sic erit in corporibus post resurrectionem. Aequalitas enim elementorum in corporibus illis excludit corruptionem in aeternum. Nam haec aequalitas est ultimus finis materiae naturalis in corporibus mixtis, quia nobilissimum est, et ideo in eo quiesceret appetitus materiae, et non desideraret aliquid ultra. Corpus autem Adae non habuit elementa in plena aequalitate, et ideo fuerunt in eo actio et passio elementorum contrariorum, et per consequens deperditio, et ideo indiguit nutrimento.

5. ARTISTS AND THE ART

Et propter hoc fuit ei praeceptum, ut non comederet de fructu vitae. Sed quia elementa in eo fuerunt prope aequalitatem, ideo modica fuit in eo deperditio; et propter hoc fuit aptus ad immortalitatem quam posset consequi, si fructum ligni vitae semper comedisset. Hic enim fructus aestimatur habere elementa prope aequalitatem; et ideo potuit continuare incorruptionem in Adam, quod factum fuisset, si non peccasset. Sapientes ergo laboraverunt, ut in aliquo comestibili vel potabili reducerent elementa ad aequalitatem vel prope, et docuerunt vias ad hoc."

23. Paravicini Bagliani, "Storia della scienza," 260.

24. Rupescissa makes a similar claim, although in a less explicit form, in DQE, 16.

25. See the unattributed passage of Bacon quoted in Paravicini Bagliani, "Storia della scienza," 259: "Necesse est etiam quod sit possibilitas huius corporis equalis, quoniam corpora in resurrectione non possunt habere incorruptionem et immortalitatem nisi per hoc corpus. Deus enim sua virtute insita de pulveribus mortuorum faciet corpus equalis complexionis ex quo constituentur corpora in resurrectione ita tamen quod subito resurget quilibet homo in tali copore equato, et tam dampnandi quam glorificandi, ut ultra non possit accidere corruptio ullo modo."

26. Boniface VIII's papal bull *Detestande feritatis* (1299) contains a number of expressions of horror at the human body "torn to pieces." The bull was written to condemn the common practice of dividing corpses for interment. See Elizabeth A. R. Brown, "Death and the Human Body in the Later Middle Ages: The Legislation of Boniface VIII on the Division of the Corpse," *Viator* 12 (1981): 221–70. For more on the general medieval anxiety about the fragmentation of the body, see Caroline Walker Bynum, *The Resurrection of the Body*.

27. Bacon, *Opus tertium*, 208; OM, 1:188. See my discussion of astrology and the prediction of Antichrist below, in chapter seven.

28. OM, 1:268–69: "Nolo hic ponere os meum in coelum, sed scio quod si ecclesia vellet revolvere textum sacrum et prophetias sacras, atque prophetias Sibyllae, et Merlini et Aquilae, et Sestonis, Joachim et multorum aliorum, insuper historias et libros philosophorum, atque juberet considerari vias astronomiae, inveniretur sufficiens suspicio vel magis certitudo de tempore Antichristi."

29. OM, 1:175: "necesse est theologum scire res hujus mundi, si textum sacrum debet scire."

30. OM, 2:221: "haec enim praecipit ut fiant instrumenta mirabilia, et factis utitur, et etiam cogitat omnia secreta propter utilitates reipublicae et personarum; et imperat aliis scientiis, sicut ancillis suis, et ideo tota sapientiae speculativae potestas isti scientiae specialiter attribuitur. Et jam ex istis scientiis tribus patet mirabilis utilitas in hoc mundo pro ecclesia Dei contra inimicos fidei, destruendos magis per opera sapientiae, quam per arma bellica pugnatorum." See also Roger Bacon, *Part of the Opus tertium of Roger Bacon, Including a Fragment Now Printed for the First Time*, ed. A.G. Little (Aberdeen: University Press, 1912), 16–17, 53–54. For a discussion of *scientia experimentalis*, see Jeremiah Hackett, "Roger Bacon on *Scientia Experimentalis*," in *Roger Bacon and the Sciences: Commemorative Essays*, ed. Hackett (Leiden: Brill, 1997), 277–315.

31. Bacon, *Opus tertium*, ed. Little, 53–54.

32. OM, 2:221–22: "quibus Antichristus copiose et efficaciter utetur, ut omnem hujus mundi potentiam conterat et confundat. . . . Et hoc deberet ecclesia considerare contra infideles et rebelles, ut parcatur sanguini Christiano, et maxime propter futura pericula in temporibus Antichristi, quibus cum Dei gratia facile esset obviare, si praelati et principes studium promoverent et secreta naturae et artis indagarent."

33. Easton, *Roger Bacon and His Search for a Universal Science*, 134–43; Marjorie Reeves, *The Influence of Prophecy in the Later Middle Ages: A Study in Joachimism* (Oxford: Clarendon Press, 1969), 48n. 1, 399; E. Randolph Daniel, *The Franciscan Concept of Mission in the High Middle Ages* (Lexington: University Press of Kentucky, 1975), 76.

34. David C. Lindberg, *Roger Bacon's Philosophy of Nature: A Critical Edition, with English Translation, Introduction, and Notes of* De multiplicatione specierum *and* De speculis comburentibus (Oxford: Clarendon Press, 1983), xxiv. See also Hilary M. Carey, "Astrology and Antichrist in the Later Middle Ages," in *Time and Eternity: The Medieval Discourse*, ed. G. Jaritz and G. Moreno-Riano (Turnhout: Brepols, 2003), 515–35; Mark T. Abate, "Roger Bacon and the Rage of Antichrist: The Apocalypse of a Thirteenth-Century Natural Philosopher," Ph.D. diss., Boston University, 1996.

35. For the sizeable bibliography on Arnald, see J. Mensa i Valls, *Arnau de Vilanova, espiritual: Guia bibliogràfica* (Barcelona: Institut d'Estudis Catalans, 1994). For an overview of Vilanova's career, see Juan A. Paniagua, *El maestro Arnau de Vilanova, médico* (Valencia: Catédra e Instituto de Historia de la Medicina, 1969); Francesco Santi, *Arnau de Vilanova: L'obra spiritual* (Valencia: Diputació Provincial de València, 1987); Arnald of Vilanova, *Arnaldi de Villanova Opera medica omnia*, ed. Michael McVaugh et al. (Granada: Seminarium Historiae Medicae Granatensis, 1981–); René Verrier, *Études sur Arnaud de Villeneuve v. 1240–1311*, 2 vols. (Leiden: Brill, 1947); Joseph Ziegler, *Medicine and Religion, c. 1300: The Case of Arnau of Vilanova* (Oxford: Clarendon Press, 1998).

36. Harold Lee, Marjorie Reeves, and Giulio Silano, *Western Mediterranean Prophecy: The School of Joachim of Fiore and the Fourteenth-Century Breviloquium* (Toronto: Pontifical Institute of Mediaeval Studies, 1989), 28.

37. Both of these were manuals designed to guide Beguin communities. They focused on the renunciation of wealth and the impending appearance of Antichrist. On the connection between the Spirituals and Arnald, see Robert E. Lerner, "Writing and Resistance Among Beguins of Languedoc and Catalonia," in *Heresy and Literacy, 1000–1530*, ed. Peter Biller and Anne Hudson (Cambridge: Cambridge University Press, 1994), 191–96; Clifford R. Backman, "The Reception of Arnau de Vilanova's Religious Ideas," in *Christendom and Its Discontents: Exclusion, Persecution, and Rebellion, 1000–1500*, ed. Scott L. Waugh and Peter D. Diehl (Cambridge: Cambridge University Press, 1995), 112–31.

38. On this incident, see Robert E. Lerner, "The Pope and the Doctor," *Yale Review* 78 (1988–89): 62–79.

39. See Juan A. Paniagua, "Notas en torno a los escritos de alquimia atribuidos a Arnau de Vilanova," *Archivo iberoamericano de historia de la medicina* 11 (1959): 404–19. See,

for instance, (Pseudo-)Arnald of Vilanova, *Le Rosier alchimique de Montpellier (Lo Rosari): Textes, traduction, notes, et commentaires*, ed. Antoine Calvet (Paris: Presses de l'Université de Paris-Sorbonne, 1997); Antoine Calvet, "Mutations de l'alchimie medicale au XVe siècle: a propos des textes authentiques et apocryphes d'Arnaud de Villeneuve," *Micrologus* 3 (1995): 185–209; Calvet, "L'alchimie d'Arnaud de Villeneuve," in *Terres médiévales*, ed. Bernard Ribémont (Paris: Klincksieck, 1993), 21–33. See also Lynn Thorndike, *A History of Magic and Experimental Science*, 8 vols. (New York: Columbia University Press, 1923–58), 3:52–84; Robert Halleux, *Les textes alchimiques*, Typologie des sources du moyen âge occidental 32 (Turnhout: Brepols, 1979), 105–6; and Calvet's article on the Pseudo-Arnaldian alchemical corpus and its relationship to both Joachism and the Franciscan Spiritual conflict, "Alchimie et joachimisme dans les alchimica pseudo-arnaldiens," in *Alchimie et philosophie à la Renaissance*, ed. Jean-Claude Margolin and Sylvain Matton (Paris: Vrin, 1993), 93–107.

40. Ziegler, *Medicine and Religion*, 40. For an extensive discussion of the authenticity of the scientific works of Arnald, see vol. 14 of the *Arxiu de Textos Catalans Antics* (1995), which is entirely devoted to this question. See especially Michela Pereira, "Arnaldo da Villanova e l'alchimia," *Arxiu de textos catalans antics* 14 (1995): 95–174.

41. Paravicini Bagliani, *The Pope's Body*, 228–29.

42. See Joseph Ziegler, "Alchemy in *Practica Summaria*: A Footnote to Michael McVaugh's Contribution," *Arxiu de textos catalans antics* 23/24 (2005): 265–67.

43. Ziegler, "Alchemy in *Practica Summaria*." See also Nicolas Weill-Parot, "Arnaud de Villeneuve et les relations possibles entre le sceau du Lion et l'alchimie," *Arxiu de textos catalans antics* 23/24 (2004–2005): 275.

44. Weill-Parot, "Arnaud de Villeneuve," 275.

45. Clifford R. Backman, "Arnau de Vilanova and the Franciscan Spirituals in Sicily," *Franciscan Studies* 50 (1990): 13–14; Backman, *The Decline and Fall of Medieval Sicily: Politics, Religion, and Economy in the Reign of Frederick III, 1296–1337* (Cambridge: Cambridge University Press, 1995), 200–209.

46. Josep Perarnau i Espelt, "*L'Allocutio christini* d'Arnau de Vilanova: edició i estudi del text," *Arxiu de Textos Catalans Antics* 11 (1992): 78–80: "Rationem vero dedit ei, ut a sensibilibus ad intelligibilia ratiocinando, sciat Dei excellentias sive dignitates animadvertere per ea, que in sensibilibus experitur, ut sic, per congnitionem ipsius in Se, quantum possibile est in presenti vita, et per cognitionem suarum dignitatum, incalescat eius animus ad amandum eum et amando sollicitetur ad laudandum eundum. Cognoscit autem homo Deum in presenti vita, primo per creaturas. In quibus, consideratis secundum originem et multitudinem et magnitudinem et pulcritudinem et ordinem et operationem, relucent ista, scilicet: potentia immensurabilis, sapientia inexplicabilis, bonitas interminabilis Creatoris."

47. Michael R. McVaugh, "Moments of Inflection: The Careers of Arnau de Vilanova," in *Religion and Medicine in the Middle Ages*, ed. Peter Biller and Joseph Ziegler (York: York Medieval Press, 2001), 50–51. Arnald did not at first include medicine among the paths to divine truth; instead, its function was confined to the advancement

of physical health. But he later came to believe that the physician's vocation allowed him to treat the soul as well as the body. William J. Courtenay has documented the substantial number of medical doctors who pursued degrees in theology in the fourteenth century. See his "Curers of Body and Soul: Medical Doctors as Theologians," in *Religion and Medicine in the Middle Ages*, ed. Peter Biller and Joseph Ziegler (York: York Medieval Press, 2001), 69–75.

48. See Ziegler, *Medicine and Religion*, 269.

49. On the printing tradition and the Pseudo-Arnaldian alchemical corpus, see Sebastià Giralt Soler, "Un alquimista medieval per als temps moderns: Les edicions del corpus alquímic atribuït a Arnau de Vilanova en llur context (c. 1477–1754)," *Arxiu de textos catalans antics* 23/24 (2005): 61–128.

50. Calvet, introduction to (Pseudo-)Arnald of Vilanova, *Le Rosier alchimique*, xxxi–xxxii.

51. Arnald of Vilanova, *Rosarius*, in *Bibliotheca chemica curiosa*, ed. Jean-Jacques Manget (hereafter cited as BCC), 2 vols. (Geneva: Chouet et al., 1702), 1:676: "conservat sanitatem: roborat firmitatem et virtutem: et de sene facit juverem, et omnem expellit aegritudinem.... Haec medicina super omnes medicinas, et mundi divitias est oppido perquirenda: quia qui habet ipsam, habet incomparabile thesaurum." A similar statement appears in the Lullian *Testamentum*, which is published as Ramón Llull, *Il "Testamentum" alchemico attribuito a Raimondo Lullo: Edizione del testo latino e catalano dal manoscritto Oxford, Corpus Christi College, 244*, ed. Michela Pereira and Barbara Spaggiari (Florence: Sismel, 1999), 378.

52. For the dating of *De vinis*, see Arnald of Vilanova, *The Earliest Printed Book on Wine*, ed. Henry E. Sigerist (New York: Schuman's, 1943), 12. For its attribution, see Calvet, introduction to (Pseudo-)Arnald of Vilanova, *Le Rosier alchimique*, xxxv; Chiara Crisciani and Michela Pereira, "Black Death and Golden Remedies: Some Remarks on Alchemy and the Plague," in *The Regulation of Evil: Social and Cultural Attitudes to Epidemics in the Late Middle Ages*, ed. Agostino Paravicini Bagliani and Francesco Santi (Florence: Sismel, 1998), 16n. 2; Michael McVaugh, "Chemical Medicine in the Medical Writings of Arnau de Vilanova," *Arxiu de textos catalans antics* 23/24 (2005): 256–64. Samantha Kelly offers evidence that a friendship between Arnald and Robert of Naples was highly unlikely. See her *The New Solomon: Robert of Naples (1309–1343) and Fourteenth-Century Kingship* (Leiden: Brill, 2003), 75n. 5.

53. Arnald of Vilanova, *Tractatus de vinis* (Paris: Felix Baligault pour Claude Jaumar et Thomas Julian, 1500), A, v. "Et usus eius preseruat proprie ab antrace et ab huiusmodi malis pustulis consumit flegma et melancoliam: et confortat proprie sustantiam cordis: et hoc facit acquirere ac consequi iuuenescentiam. Et forte qui assiduauerit eo non putrefiet corpus eius."

54. Mentions of wine as a remedy for the plague appear in medical literature during the time of the plague, for example, in the *consilia* of Gentile da Foligno. In his third *consilium*, he recommends theriac, leeks, scallions, and "bitter herbs" either mixed with or boiled in wine. An excerpt from Gentile's plague *consilia* appears in John Aberth,

5. ARTISTS AND THE ART

The Black Death: The Great Mortality of 1348–1350, a Brief History with Documents (New York: Palgrave, 2005), 49–50.

55. The MS copy of *De aqua vite* is Cambrai 919 (818) Lat. fols. 145v–161v. Calvet notes that this is the same work as the printed Arnald of Vilanova, *Clarissimi et excellentissimi philosophi medicique magistri Arnaldi de villanova de aqua vite simplici et composita tractatus pulcherimus incipit* (Venice: Adam de Rottweill, c. 1477–8 [s.l.n.d.]). See Calvet, "Alchimie et joachimisme," 102–1. Also see his introduction to (Pseudo-)Arnald of Vilanova, *Le Rosier alchimique*, xxxvi.

56. Rupescissa's description of the *hierapicra* and its relationship to the astrological sign of Aries leads Halleux to conclude that *De quinta essentia* was inspired by similar theories in *De aqua vite*. See his "Ouvrages," 253.

57. DQE, 29.

58. Although it is possible that this is the origin of Rupescissa's claim in the *Liber lucis* that a "quintessence" was involved in transmutation. See Arnald of Vilanova, *Quaestiones tam essentiales quam accidentales*, in BCC, 701.

59. Arnald of Vilanova, *Testamentum*, in BCC, 704.

60. Ziegler, *Medicine and Religion*, 54; J. N. Hillgarth, *Ramon Lull and Lullism in Fourteenth-Century France* (Oxford: Clarendon Press, 1971), 53–56, 97, 261.

61. See Lola Badía and Anthony Bonner, *Ramón Llull: Vida, pensamiento y obra literaria* (Barcelona: Sirmio, 1992); Ramón Llull, *Selected Works*, 2 vols., ed. and trans. Anthony Bonner (Princeton, N.J.: Princeton University Press, 1985); Frances A. Yates, "The Art of Ramon Lull," *Journal of the Warburg and Courtald Institutes* 17 (1954): 115–73; Mark D. Johnston, *The Evangelical Rhetoric of Ramon Llull: Lay Learning and Piety in the Christian West Around 1300* (New York: Oxford University Press, 1996).

62. Yates, "The Art of Ramon Lull," 116–17.

63. Jean de Roquetaillade (John of Rupescissa), *Liber ostensor quod adesse festinant tempora*, ed. Clémence Thévenaz Modestin, Christine Morerod-Fattebert, André Vauchez, et al. (Rome: École Française de Rome, 2005) (cited hereafter as LO), 830: "Et quia tempore Celestini quinti et ultra, per totum tempus *regis coronatorum*, predicti Philippi Pulcri, floruit philosophus barbatus—sic vocatus—Parisius, qui dicebatur Ramundus Luylh de Majoricis, qui, cum esset laycus et quasi sine literis, ad tantum apicem philosophie pervenit ut XXXV libros inventivos omnium scienciarum et solutivos omnium questionum tam stupendos fecerit quod mens humana credere non valeret, nisi eorum experienciam cum intellectu videret; fecitque plures alios, circa ducentos inter omnes."

64. Ziegler, *Medicine and Religion*, 54–58.

65. Ramón Llull, *Liber principiorum medicine*, in *Selected Works*, 2:1157–61, 1199–1207. The pepper analogy is from the *Arbor christianalis*. See Yates, "The Art of Ramon Lull," 152.

66. Badía and Bonner, *Ramón Llull*, 62; see Rupescissa's complaints of being called *fantasticus* in LO, 513–16.

5. ARTISTS AND THE ART

67. Michela Pereira has written a number of studies on these texts. See, for instance, *The Alchemical Corpus Attributed to Raymund Llull* (London: Warburg Institute, 1989); Pereira, "*Medicina*," 1–15; and Llull, *Il "Testamentum" alchemico*; See also William R. Newman's corpuscular interpretation of the *Testamentum* in *Gehennical Fire: The Lives of George Starkey, an American Alchemist in the Scientific Revolution* (Cambridge, Mass.: Harvard University Press, 1994), 99–103.

68. Llull, *Il "Testamentum" alchemico*, 306, 378. See Pereira, "*Medicina*," 6.

69. Llull, *Il "Testamentum" alchemico*, 12–16.

70. For a detailed analysis of the *Testamentum*, see Pereira, *Alchemical Corpus*, 14. On *De generatione stellarum*, see chapter 4, page 75.

71. Pereira, *Alchemical Corpus*, 10–16.

72. Other versions have additional books; see Pereira, *Alchemical Corpus*, 12.

73. (Pseudo-)Llull, *Raimundi Lulii Maiorici, philosophi acutissimi, medicique celeberrimi, De secretis naturae siue quinta essentia libri duo* (Venice: Petrus Schoeffer, 1542), 84–85.

74. Pereira, "*Medicina*," 2.

75. Halleux, "Ouvrages," 273; Pereira, *Alchemical Corpus*, 11–19.

6. METAPHOR AND ALCHEMY

1. For an introduction to *Decknamen*, see Robert Halleux, *Les textes alchimiques*, Typologie des sources du moyen âge occidental 32 (Turnhout: Brepols, 1979), 34–35, 114–19; Halleux, "Problèmes de lexicographie alchimiste," in *La lexicographie du latin médiéval et ses rapports avec les recherches actuelles sur la civilisation du moyen âge*, ed. Yves Lefèvre (Paris: CNRS, 1981), 355–65; and also see William R. Newman, "'Decknamen or Pseudochemical Language'? Eirenaeus Philalethes and Carl Jung," *Revue d'histoire des sciences* 49 (1996): 159–88; William R. Newman and Anthony Grafton, "Introduction: The Problematic Status of Astrology and Alchemy in Premodern Europe," and William R. Newman and Lawrence M. Principe, "Some Problems with the Historiography of Alchemy," both in *Secrets of Nature: Astrology and Alchemy in Early Modern Europe*, ed. Newman and Grafton (Cambridge, Mass.: MIT Press, 2001), 1–37, 385–431. Several scholars have usefully examined the problem of alchemical imagery from art historical, literary, and theological viewpoints. See, for instance, Barbara Obrist, *Les débuts de l'imagerie alchimique (XIVe-XVe siècles)* (Paris: Sycomore, 1982); Lee Patterson, "Perpetual Motion: Alchemy and the Technology of the Self," *Studies in the Age of Chaucer* 15 (1993): 25–57; Barbara Newman, *God and the Goddesses: Vision, Poetry, and Belief in the Middle Ages* (Philadelphia: University of Pennsylvania Press, 2003), 234–43.

2. See, for instance, William R. Newman's otherwise excellent reading of the fourteenth-century *Testamentum*, attributed to Raymond Llull, in *Gehennical Fire: The Lives of George Starkey, an American Alchemist in the Scientific Revolution* (Cambridge, Mass.: Harvard University Press, 1994), 98–99, 115–16. Lawrence Principe offers a similar analysis in *The Aspiring Adept: Robert Boyle and His Alchemical Quest, Including Boyle's "Lost" Dialogue on the Transmutation of Metals* (Princeton, N.J.: Princeton University Press, 1998), 188–90.

6. METAPHOR AND ALCHEMY

3. See Theodore L. Brown, *Making Truth: Metaphor in Science* (Urbana: University of Illinois Press, 2003); George Lakoff and Mark Johnson, *Metaphors We Live By* (Chicago: University of Chicago Press, 1980); Earl R. MacCormac, *Metaphor and Myth in Science and Religion* (Durham, N.C.: Duke University Press, 1976); and the essays collected in Andrew Ortony, ed., *Metaphor and Thought* (Cambridge: Cambridge University Press, 1993), especially Thomas S. Kuhn, "Metaphor and Science," 533–42, and Dedre Gentner and Michael Jeziorski, "The Shift from Metaphor to Analogy in Western Science," 447–80.

4. John of Rupescissa, *Liber de consideratione quinte essentie omnium rerum* (Basel, 1561) (hereafter cited as DQE), 25, 28, 50–51.

5. Barbara Obrist, "Les rapports d'analogie entre philosophie et alchimie médiévales," in *Alchimie et philosophie à la Renaissance*, ed. Jean-Claude Margolin and Sylvain Matton (Paris: Vrin, 1993), 49; Albertus Magnus, *Book of Minerals*, trans. Dorothy Wyckoff (Oxford: Clarendon Press, 1967), 204 (4.1), 21–23 (1.1.5).

6. These two stages are described by Obrist, "Les rapports," 43–64. Obrist does an excellent job of analyzing the religious context of Rupescissa's alchemy in her *Débuts*, 58, 138–40.

7. Obrist, "Les rapports," 54–64. Obrist links this new reliance on religious faith to the ambiguity of alchemy's ability to transmute metals. Chiara Crisciani makes a similar argument in "The Conception of Alchemy as Expressed in the *Pretiosa Margarita Novella* of Petrus Bonus of Ferrara," *Ambix* 20 (1973): 170. On secrecy in alchemical literature, see William Eamon, *Science and the Secrets of Nature: Books of Secrets in Medieval and Early Modern Culture* (Princeton, N.J.: Princeton University Press, 1994), 11, 76–90; Karma Lochrie, *Covert Operations: The Medieval Uses of Secrecy* (Philadelphia: University of Pennsylvania Press, 1999), 93–118, Pamela O. Long, *Openness, Secrecy, Authorship: Technical Arts and the Culture of Knowledge from Antiquity to the Renaissance* (Baltimore, Md.: Johns Hopkins University Press, 2001).

8. Newman and Principe, "Some Problems with the Historiography of Alchemy," 388, 404–15. See also Luther H. Martin, "A History of the Psychological Interpretation of Alchemy," *Ambix* 22 (1975): 10–20.

9. Robert P. Multhauf, "John of Rupescissa and the Origin of Medical Chemistry," *Isis* 45 (1954): 364; Robert Halleux, "Les ouvrages alchimiques de Jean de Roquetaillade," *Histoire littéraire de la France* 41 (1981): 252.

10. DQE, 21: "Quod autem incorruptibilitati conferrat, et a corruptibilitate praeservet, demonstrabo ex experientia assumpta: quoniam si quaecunque avis, aut carnis frustum, aut piscis infundatur in ea, non corrumpetur quandiu permanebit in ea: quanto magis ergo carnem animatam et vivam corporis nostri ab omni corruptione servabit? Haec quinta Essentia est caelum humanum, quod creavit Altissimus ad conservationem quatuor qualitatum corporis humani, sicut caelum ad conservationem totius universi. Et scito pro certo quod Philosophi hodierni et medici ignorant omnino hanc quintam Essentiam et veritatem et virtutem eius: sed ego, auxilio Dei, inferius revelabo tibi magisterium eius: et huiusque docui te rem secretam, quintam Essentiam, id est caelum humanum."

{205}

6. METAPHOR AND ALCHEMY

11. See chapter 4, page 68.

12. DQE, 22: "Et sicut caelum summum non influit solum per se conservationem in mundo et influentias miras, sed per virtutem solis et aliarum stellarum: sic et caelum istud, quinta Essentia vult ornari sole mirabili . . . qui est creatus ad ornatum caeli nostri, et ad augendum influentiam quintae Essentiae."

13. As in Multhauf's and Frank Sherwood Taylor's brief accounts of Rupescissa's quintessence; see Multhauf, "Origin," 364; Taylor, "The Idea of the Quintessence," in *Science, Medicine, and History: Essays on the Evolution of Scientific Thought and Medical Practice written in Honour of Charles Singer*, ed. E. Ashworth Underwood (London: Oxford University Press, 1953), 258–59.

14. On the nature of the resurrected body, see Caroline Walker Bynum, *The Resurrection of the Body in Western Christianity, 200–1336* (New York: Columbia University Press, 1995). On heaven, see Paul Binski, *Medieval Death: Ritual and Representation* (Ithaca, N.Y.: Cornell University Press, 1996), 166–81, 199–203; Eileen Gardiner, *Visions of Heaven and Hell Before Dante* (New York: Italica Press, 1989).

15. W. Ganzenmüller, *Die Alchemie im Mittelalter* (Paderborn: Verlag der Bonifaciusdruckerei, 1938), was the first major study of Christocentric imagery in alchemical texts, especially those of Arnald of Vilanova, Petrus Bonus, and Rupescissa. More recently, Antoine Calvet and Chiara Crisciani have further analyzed Christocentric imagery in the alchemy of, respectively, Arnald of Vilanova and Petrus Bonus. See, for instance, (Pseudo-)Arnald of Vilanova, *Le Rosier alchimique de Montpellier (Lo Rosari): Textes, traduction, notes, et commentaires*, ed. Antoine Calvet (Paris: Presses de l'Université de Paris-Sorbonne, 1997); Calvet, "Alchimie et joachimisme dans les alchimica pseudo-arnaldiens," in *Alchimie et philosophie à la Renaissance*, ed. Jean-Claude Margolin and Sylvain Matton (Paris: Vrin, 1993), 93–107; Petrus Bonus of Ferrara (Pietro Bono da Ferrara), *Preziosa margarita novella: Edizione del volgarizzamento, introduzione e note*, ed. Chiara Crisciani (Florence: Nuova Italia Editrice, 1976); Crisciani, "The Conception of Alchemy," 165–81. See also Michela Pereira, *The Alchemical Corpus Attributed to Raymund Llull* (London: Warburg Institute, 1989).

16. Petrus Bonus of Ferrara, *Pretiosa margarita novella*, in in *Bibliotheca chemica curiosa*, ed. Jean-Jacques Manget (hereafter cited as BCC), 2 vols. (Geneva: Chouet et al., 1702), 2:30: "tunc similis videtur mortuo, et tunc res illa igne indiget, quousque illius corporis spiritus extrahatur . . . et pulvis fiat. Iis peractis, Deus reddet ei animam suam et spiritum , ac infirmitate ablata, confortatur illa res, et post coruscationem emendatur, quemadmodum homo post resurrectionem. . . . unde est conceptio similis conceptione virginis, quae absque viro concipit: quod esse non potest nisi miraculose, scilicet per divinam gratiam."

17. Crisciani, "The Conception of Alchemy," 171–73.

18. Calvet, "Alchimie et joachimisme," 99–102. See also the excerpt from the *Exempla philosophorum*, a work attributed to Arnald, in Chiara Crisciani and Michela Pereira's *L'arte del sole e della luna: alchimia e filosofia nel medioevo* (Spoleto: Centro italiano di studi sull'alto medioevo, 1996), 233–39.

19. Lynn Thorndike, *A History of Magic and Experimental Science*, 8 vols. (New York: Columbia University Press, 1923–58), 3:75–78.

6. METAPHOR AND ALCHEMY

20. Rupescissa quotes or paraphrases "Magister Arnaldus" or "Arnaldus de Vilanova" seven times in the short treatise; four of these citations deal with the philosophers' stone as Christ. See John of Rupecissa, *Liber lucis*, in *Il libro della luce*, ed. Andrea Aromatico (Marsilio: Editori, 1997) (hereafter cited as LL), fol. 20v, 21v, 23r, 24v, 25r, 28r.

21. LL, 25r–v: "scito autem et animadverte quod multi Magistri, quibus revelatum fuit hunc misterium a revelantibus, imprecati fuerunt et maledicti horrendus maledictiones." On Hermes Trismegistus and Alphidius, see Halleux, "Ouvrages," 266; Halleux, *Les textes alchimiques*, 84–85.

22. Newman, *Gehennical Fire*, 117.

23. See Calvet's remarks in (Pseudo-)Arnald of Vilanova, *Le Rosier alchimique*, xxxi. He notes that part of this passage is common to the *Rosarius philosophorum* and the *Liber lucis*. See also Barthélemy Haureau, "Arnauld de Villeneuve, médecin et chimiste," *Histoire littéraire de la France* 28 (1881): 92.

24. LL, 20v: "Rainaldus de Villanova inquit in tractatu parabolico de Maiori Edicto, sub tipo verborum Evangelii et crucifixionis Iesu Christi, illa verba: 'nisi granum frumenti cadens in terra mortuum fuerit, ipsum solum manet; si autem mortuum fuerit, multum fructum affert.' Intelligens per granum frumenti Mercurium Vivum mortificatum in terra Salis Petre et Vitrioli Romani, et ibi mortificatum cum igne in sublimatione multum fructum affert, quia est Lapis magnis quem Philosophi quesierunt."

25. See Betty Jo Teeter Dobbs's discussion of medieval alchemy in her *Alchemical Death and Resurrection: The Significance of Alchemy in the Age of Newton* (Washington, D.C.: Smithsonian Institution Libraries, 1990), 13–15.

26. Caroline Walker Bynum, "Metamorphosis, or Gerald and the Werewolf," in *Metamorphosis and Identity* (New York: Zone Books, 2001), 85.

27. LL, 24r–v: "et ignem predictum continua donec videbis materiam denigrari, quod, si nimis tarderetur, augeatur ei ignis et, cum videris materiam denigrari, gaude quia tunc habes principium digestionis, et tunc continua ignem donec omnes alii colores transeant et videbis aliquantulum dealbari et tunc augmenta ei ignem paulatim et insensibiliter quod videris magis ac magis materiam dealbari et, cum videris materiam dealbatam, est perfecta et tunc lapidem habes perfectum ad album, convertens omnia metalla et omnia corpora metallica in perfectum ac melius argentum quam sit de minera."

28. On the necessity of *nigredo* for generation, see Noel L. Brann, "Alchemy and Melancholy in Mediaeval and Renaissance Thought: A Query Into the Mystical Basis of their Relationship," *Ambix* 32 (1985): 135–37.

29. As quoted in Brann, "Alchemy and Melancholy," 136, using the edition of the *Speculum alchimiae* in Lazarus Zetzner, *Theatrum chemicum* (Strassburg, 1613–1661), 4:607. Calvet's edition of the Arnald's *Rosarius philosophorum* contains a similar passage: "D'où aussitôt que la forme propre du corps est dissoute, aussitôt une autre forme nouvelle entre dans ce corps, une forme corrompue, de couleur noire, d'une odeur puante et au contact fort subtile et discontinue. Voilà les signes de la parfaite dissolution du corps" ([Pseudo-]Arnald of Vilanova, *Le Rosier alchimique*, 7).

6. METAPHOR AND ALCHEMY

30. Dobbs, *Alchemical Death and Resurrection*, 15.

31. Brann charts this same sequence of color changes in the alchemy of Pseudo-Llull in "Alchemy and Melancholy," 135.

32. Psalm 117:22; Isaiah 8:13–14; Dobbs, *Alchemical Death and Resurrection*, 12.

33. Acts 4:11; 1 Peter 2:4–8; Dobbs, *Alchemical Death and Resurrection*, 12.

34. LL, 23r: "Nam de hac quarta operatione dicitur expressio Lacte Virginis, unde Magister Arnaldus dicit: "quod oportet a terra exaltari filius hominis in aeris cruce" que ad litteram de materia in tertia operatione digesta, ubi ponitur Mercurius ad dissolvendum in fundo urinalis, et quod ascendit ibi purum est et spirituale et in aerem farinosum convertitur et exaltatur in cruce capitis alembicci quasi Christus, ut Magister Arnaldus dicit."

35. LL, 27r: "Deinde fac fieri laminam ferream fortem et rotundam, habentem quatuor brachia ad modum crucis."

36. John of Rupecissa, *Liber secretorum eventuum*, ed. Robert E. Lerner and Christine Morerod-Fattebert (Fribourg, Switzerland: Editions Universitaires, 1994), 109–10: "Item ibidem, in prefato capitulo II, dicit textus quod *lapis qui abscisus est de monte sine manibus percussit statuam in pedibus eius ferreis et fictilibus et comminuit eos.* Evidens enim est quod pedes lutei, in quibus percutitur statua a lapide, sunt tempora et tyranni temporum Antichristi et finis Imperii Romani dividendi in decem digitos. . . . Futurum ergo est iuxta sensum principalem ut Romanum Imperium tempore decem digitorum—hoc est decem regum—fictilium a lapide Christo percutiatur in Antichristo et ad nichilum redigatur."

37. A similar use of the Christ-stone analogy appears in Rupescissa's sixth treatise of the commentary on the *Oraculum Cyrilli*, in which Rupescissa describes Christ as a stone, an identification that he attributes to St. Paul. Jeanne Bignami-Odier, "Jean de Roquetaillade (de Rupescissa), Théologien, Polémiste, Alchimiste," *Histoire littéraire de la France* 41 (1981): 115; Halleux, "Ouvrages," 245n. 32.

38. John of Rupescissa, *Vade mecum in tribulatione*, Venice, Biblioteca Nazionale Marciana, MS Lat. III, 39, fol. 1–19v (hereafter cited as VM), int 9: "ipse est post christum qui principio tertii status orbis generalis lapis abscissus de monte sine manibus qui babillonicam statuam percutiet in pedibus fictilibus et eam cominuet et implebit lapidem christum ac eius legem universam terram omnes infideles exceptis angulis orbis a ghogh et magogh qui remanebunt convertendo ad dominum de isto tractatu prophete ysa yer et caeteri ubicunque agitur de ipso fere que secundo sensu pro certo statu scilicet memorato nisi quia ea que ad divinitatem christo conveniunt debent duntaxat reservari domino jesu christo."

39. Petrus Bonus of Ferrara, *Pretiosa margarita novella*, in BCC, 2:30–31, 34; Crisciani, "The Conception of Alchemy," 171–72.

40. Eric John Holmyard, *Alchemy* (Harmondsworth: Penguin, 1957), 16; Stanton J. Linden, *Darke Hierogliphicks: Alchemy in English Literature from Chaucer to the Restoration* (Lexington: University of Kentucky Press, 1996), 8.

41. Linden, *Darke Hierogliphicks*, 10.

6. METAPHOR AND ALCHEMY

42. Rupescissa's antimony is actually antimony sulphide, the ore of metallic antimony. Robert Multhauf has speculated that the red liquid is antimony trisulphide, which was later known as Kermes mineral. See Multhauf, "Origin," 365; Multhauf, *The Origins of Chemistry* (1966; reprint, Langhorne, Penn.: Gordon and Breach Science Publishers, 1993);, 211. Red-colored chemicals as agents of longevity also appear frequently in Chinese alchemy. See Gerald J. Gruman, *A History of Ideas About the Prolongation of Life: The Evolution of Prolongevity Hypotheses to 1800*, Transactions of the American Philosophical Society 56:9 (Philadelphia: American Philosophical Society, 1966), 51–52.

43. DQE, 104–5: "deinde videbis miraculum stupendum, quoniam per rostrum alembici videbis quasi mille venulas liquorum benedictae minerae descendere per guttas rubeas sicut recte esset sanguis ... aufert dolorem omnium vulnerum et mirabiliter sanat. Virtus eius est incorruptibilis et miraculosa et utilis supra modum."

44. LL, 24r: "Et dixerunt quidam Philosophi; "animal cum suo sanguine." Animal quia crescit sublimando et quia habet animam sanguinis id est Tincturam Rubedinis cum spiritu Vitrioli Romani supradicto." This passage seems to invoke Pseudo-Avicenna's *De arte in alkimia*, which refers to an "animal stone" that could be interpreted as human blood. See William R. Newman, "The Philosopher's Egg: Theory and Practice in the Alchemy of Roger Bacon," *Micrologus* 3 (1995): 82. Rupescissa may also have in mind Leviticus 17:11: "Quia carnis anima in sanguine est et ego dedi illum vobis ut super altare in eo expietis pro animabus vestris et sanguis pro animae piaculo sit [Because the soul of the flesh is in the blood: and I have given it to you, that you may make atonement with it upon the altar for your souls, and the blood may be for an expiation of the soul]."

45. LL, 24v: "Sic extrahe ipsum de vase predicto quod vocatur ovum Philosophorum, et Magister Arnaldus dicit quod Lapis est clausus in eo sicut Christus in sepulcro."

46. LL, 25r: "et ita nichilque remoto inde Lapis albus per maiorem ignem efficitur rubeus, et ascendet de sepulchro Rex Excellentissimus a mortuis resuscitatus et cum diademate rutilante, fulgens et gloriosus, ut Magister Arnaldus testatur."

47. Newman, "The Philosopher's Egg," 80. Roger Bacon, *Secretum secretorum cum glossis et notulis*, ed. Robert Steele, vol. 5 of *Opera hactenus inedita Rogeri Baconi* (Oxford: Clarendon Press, 1909), 117: "Lapis igitur sumitur primo methaphorice pro omni eo super quo incipit operacio alkimie. Et hoc potest esse res mineralis, ut sulphur et arsenicum, set melior est res vegetabilis ut fructus et partes arborum et herbarum, optime vero sunt res animales ut sanguis ovum et capilli, et maxime partes hominis, et inter illas sanguis, in quo ad oculum distinguntur quatuor humores, scilicet, fluema, colera, sanguis, et melancolia." ("'Stone' is first taken metaphorically for any thing upon which an alchemical operation begins. This can be mineral matter, such as sulfur and arsenic, but vegetable matter, such as fruits and the parts of trees and herbs is better, and best of all are the animal materials such as blood, egg, and hairs, and especially the parts of man, and among them [primarily] blood, in which the four humors, namely phlegm, choler, blood, and melancholy, are distinguished by the eye." The translation is Newman's.) See also the essays collected in W. F. Ryan and Charles B. Schmitt, eds., *Pseudo-Aristotle, The Secret of Secrets: Sources and Influence* (London: Warburg Institute, 1982).

6. METAPHOR AND ALCHEMY

48. On Pseudo-Avicenna, see Michela Pereira, "Teorie dell'elixir nell'alchimia latina medievale," *Micrologus* 3 (1995): 108–16.

49. Newman, "Philosopher's Egg," 81.

50. (Pseudo-)Arnald of Vilanova, *Epístola Magistri Arnaldi Cathelani de Villanova ad Magistrum Iacobum de Toleto de sanguine humano*, printed in same volume as *De quinta essentia* (Basel, 1561), 169–74. An excerpt is printed in Joseph Frank Payne, "Arnald de Villa Nova on the Therapeutic Use of Human Blood," *Janus* 8 (1903): 433–35. On this treatise, see Haureau, "Arnauld de Villeneuve," 92–93.

51. Halleux, "Ouvrages," 254–55.

52. Piero Camporesi includes a number of these in a chapter entitled "Quintessence of Blood," but he incorrectly traces the idea of the restorative quintessence to Llull, rather than to Rupescissa. See Camporesi, *Juice of Life: The Symbolic and Magic Significance of Blood*, trans. Robert R. Barr (New York: Continuum, 1995), 27–52. For the appropriation of Rupescissa's quintessence alchemy by the author of the Pseudo-Llullian *De secretis naturae*, see my discussion in chapter 5.

53. See his recipe for the quintessence of blood, DQE, 42–43.

54. It is also significant that concomitance—the doctrine asserting that the body and blood of Christ are fully present in every morsel of transubstantiated bread or wine—was used to justify the withholding of the eucharistic cup from the laity. According to Megivern, concomitance was an early medieval idea that was elaborated by theologians in the late eleventh century in response to an assertion by Berengar of Tours that communion divided Christ into "little pieces." The application of the doctrine to the consecrated bread and wine may further point to the significance of blood during this period. See James J. Megivern, C.M., *Concomitance and Communion: A Study in Eucharistic Doctrine and Practice* (Fribourg: University Press, 1963), 238. On blood imagery in late-medieval art and culture, see Caroline Walker Bynum, *Wonderful Blood: Theology and Practice in Late Medieval Northern Germany and Beyond* (Philadelphia: University of Pennsylvania Press, 2007); James H. Marrow, *Passion Iconography in Northern European Art of the Late Middle Ages and Early Renaissance: A Study of the Transformation of Sacred Metaphor Into Descriptive Narrative* (Kortrijk: Van Ghemmert, 1979); for devotion to Christ's blood, which began to intensify in the twelfth century, see Louis Gougaud, *Devotional and Ascetic Practices in the Middle Ages*, trans. G. C. Bateman (London: Burns, Oates, and Washbourne, 1927), 75–130; the chapter "Ideal of the Imitation of Christ," in Giles Constable, *Three Studies in Medieval Religious and Social Thought* (Cambridge: Cambridge University Press, 1995), 209–17.

55. See, for instance, Marlene Villalobos Hennessy's analysis of devotional images of Christ's wound and blood in "Aspects of Blood Piety in a Late Medieval English Manuscript: London, British Library Additional 37049," in *History in the Comic Mode: Medieval Communities and the Matter of Person*, ed. Rachel Fulton and Bruce W. Holsinger (New York: Columbia University Press, 2007), 182–91; Hennessy, "Morbid Devotions: Reading the Passion of Christ in a Late Medieval Miscellany, London, B.L. MS Additional 37049," Ph.D. diss., Columbia University, 2001. See also Bynum, *Wonderful Blood*.

56. Hennessy, "Aspects of Blood Piety" 174–76, and "Morbid Devotions," and 257.

6. METAPHOR AND ALCHEMY

57. Jeffrey F. Hamburger, *Nuns as Artists: The Visual Culture of a Medieval Convent* (Berkeley: University of California Press, 1997), 103–28.

58. Hamburger, *Nuns as Artists*, fig. 67 and plate 10.

59. Such images are related to the role of blood in sacrifice. Christ's blood spilled in sacrifice was depicted as generative but also accusatory, as I shall discuss below. On the Eucharist as a sacrifice, as well as a sacrament, see Jaroslav Pelikan, *The Growth of Medieval Theology (600–1300)*, vol. 3 of *The Christian Tradition: A History of the Development of Doctrine* (Chicago: University of Chicago Press, 1978), 188–95.

60. Hennessy, "Morbid Devotions," 175. Such healing was sometimes interpreted to benefit the body as well as the soul. For early modern opinions on the regenerative power of blood, see Camporesi, *Juice of Life*, 14, 31–32.

61. Bynum, *Wonderful Blood*, 47–68, 155–56.

62. DQE, 41–42: "Si audivisti quod tibi superius ex secretis maximis revelavi, cum dixi, quod in omni re quinta Essentia est et remanet incorrupta, maximum erit tibi si docuero ipsam extrahere a sanguine humano, et a carnibus omnium animalium, et ab ovis, et a similibus rebus. Quoniam cum sanguis humanus sit perfectum opus naturae in nobis quantum ad augmentum deperditum, certum est, quod natura illam quintam Essentiam sic perficit et perfecit, ut sine alia praeparatione ex venis transferat ipsum sanguinem immediate in carnem: et hanc quintam Essentiam tam propriam naturae maximum est haberi: quia in ea est virtus mirabilis caeli stellati nostri: et ad curam naturae divinissima operatur miracula, sicut infra docebo. Recipe ergo a barbutonsoribus ex omnibus iuvenibus sanguineis vel cholericis sanguinem assumptum ut poteris ab eis qui vinis bonis utuntur, et abiecta aqua sive phlegmate."

63. Nancy G. Siraisi, *Medieval and Early Renaissance Medicine: An Introduction to Knowledge and Practice* (Chicago: University of Chicago Press, 1990), 105–6; Oswei Temkin, *Galenism: Rise and Decline of a Medical Philosophy* (Ithaca, N.Y.: Cornell University Press, 1973), 17.

64. Caroline Walker Bynum, *Fragmentation and Redemption: Essays on Gender and the Human Body in Medieval Religion* (New York: Zone Books, 1992), 100, 109–14; Danielle Jacquart and Claude Thomasset, *Sexuality and Medicine in the Middle Ages*, trans. Matthew Adamson (Oxford: Polity Press, 1988), 34, 52–54. On the Aristotelian idea of blood as the basic fluid, see Anthony Preus, "Galen's Criticism of Aristotle's Conception Theory," *Journal of the History of Biology* 10 (1977): 65–85.

65. Caroline Walker Bynum, "The Blood of Christ in the Middle Ages," *Church History* 71 (2002): 685–714; Bynum, "Violent Imagery in Late Medieval Piety," *German Historical Institute Bulletin* 30 (2002): 1–34.

66. Bynum, "Violent Imagery," 25–29. This burden of guilt was often displaced onto Jews, but Christian blasphemers, criminals, or those who violated the Sabbath were also targeted. Devotional images depicted such sinners as murderers of Christ; the message was that humans continued to murder Christ every day through sin. On this element of sacrifice as accusative, see also Pelikan, *The Growth of Medieval Theology*, 129–44, 184–204.

{211}

6. METAPHOR AND ALCHEMY

67. Marie-Christine Pouchelle, *The Body and Surgery in the Middle Ages*, trans. Rosemary Morris (Oxford: Polity Press, 1990), 13, 20–21; see also Jerome Kroll and Bernard Bachrach, "Sin and the Etiology of Disease in Pre-Crusade Europe," *Journal of the History of Medicine and Allied Sciences* 41 (1986): 395–414.

68. See Jacquart and Thomasset on menstruation in *Sexuality and Medicine in the Middle Ages*, 71–78. Menstrual blood was particularly ambiguous: anecdotes of menstruating women who ruined crops or clouded mirrors were common in medical and theological discourse, as were associations of menstruation with sin. Charles T. Wood argues that certain scholastic theologians were even reluctant to affirm that the Virgin Mary menstruated in his "The Doctors' Dilemma: Sin, Salvation, and the Menstrual Cycle in Medieval Thought," *Speculum* 56, no. 4 (1981): 710–27. Where Wood sees a contradiction between the Virgin's purity and maternity, however, Paulette L'Hermite-Leclercq sees a mystery that engendered no dilemma for believers. See her "Le sang et le lait de la vierge," in *Le sang au Moyen Âge, Actes du quatrième colloque international de Montpellier* (Montpellier: C.R.I.S.I.M.A., 1999), 145–62.

69. For a theoretical approach to descriptions of violence and similar "outpourings of blood," see Mark D. Meyerson, Daniel Thiery, and Oren Falk, eds., *'A Great Effusion of Blood': Interpreting Medieval Violence* (Toronto: University of Toronto Press, 2003).

70. VM, int. 5: "fames gravissime supra mundum pestilentiales et mortalitates et gutturum sanguinatione et alie apostematice passiones quibus plagis interficietur maxima pars [p]rave generationis presentis ut renovetur orbis et superbi indurati deleantur de mundo ad veritatem reducendum."

71. Joseph Ziegler, *Medicine and Religion, c. 1300: The Case of Arnau of Vilanova* (Oxford: Clarendon Press, 1998).

72. Martin Aurell has recently analyzed Rupescissa's merciless attitude toward Muslims by grounding it within the changing political agenda of the papacy in the 1350s. He views Rupescissa's call for the destruction of Muslims as stemming from political currents within the papal court, which became decidedly pro-Crusade during this period. See Aurell, "Prophétie et messianisme politique: La péninsule iberique au miroir du *Liber ostensor* de Jean de Roquetaillade," in *Textes prophétiques et la prophétie en Occident (XIIe-XVIe siècle)*, ed. André Vauchez (Rome: École Française de Rome, 1990), 27–71.

73. See Peter Brown, *The Body and Society: Men, Women, and Sexual Renunciation in Early Christianity* (New York: Columbia University Press, 1988), 117.

74. Brown, *The Body and Society*, 140, 168–69.

75. Matthew 5:30: "et si dextera manus tua scandalizat te abscide eam et proice abs te expedit tibi ut pereat unum membrorum tuorum quam totum corpus tuum eat in gehennam."

76. Richard Kieckhefer, *Unquiet Souls: Fourteenth-Century Saints and Their Religious Milieu* (Chicago: University of Chicago Press, 1984), 124, 147–49. Interestingly, Astesano di Asti's fourteenth-century confessor's manual, *Summa de casibus conscientie*, forbids bodily mutilation to avoid sin or reduce temptation. See Ziegler, *Medicine and Religion*, 8.

7. THE END OF NATURE

77. The *Vade mecum* may also show the diverging approaches to disease of Galenic and surgical medicine. On the relationship between the two, see Pouchelle, *The Body and Surgery in the Middle Ages*.

7. THE END OF NATURE

1. John of Rupescissa, *Liber de consideratione quinte essentie omnium rerum* (Basel, 1561) (hereafter cited as DQE), 25–26: "Et scito, quod universae stellae caeli habent suam influentiam singularem ex iussu et ordinatione Dei, et quaelibet stella suam proprietatem habet super rem determinatam et certam: ut stella poli super Adamantem et ferrum: et status Lune super aquam maris, Sol super aurum: Luna super agentum: imagines hominum caeli super corpora humana: imago arietis caeli super arietes terrestres et caput hominis. Et sicut carpentarius cum dolabra in manu fabricans arcas, non minus fabricaret: sic nec Deus minus gubernat mundum: sic dedit talem influxum stellis, ut influant in rebus sicut et quomodo ipse vult, et non ultra." On the connotations of inflowing in celestial influence, see John D. North, "Celestial Influence—the Major Premiss of Astrology," in *'Astrologi hallucinati': Stars and the End of the World in Luther's Time*, ed. Paola Zambelli (Berlin: Walter de Gruyter, 1986), 45. For the mechanics of influence, see Olaf Pedersen, "Astronomy," in *Science in the Middle Ages*, ed. David C. Lindberg (Chicago: University of Chicago Press), 1978, 304–5.

2. For an analysis of the God-as-craftsman metaphor, see George Ovitt, *The Restoration of Perfection: Labor and Technology in Medieval Culture* (New Brunswick, N.J.: Rutgers University Press, 1987), 57–87. Rupescissa's assertion that celestial influence obeys *iussu Dei*—the will of God—follows closely the discussion of the ninth-century Arab astrologer Abū Ma'shar (known in Latin as Albumasar) in the *Introductorium in astronomiam*, a work of tremendous import for natural philosophers and alchemists, including Roger Bacon and Albert the Great. The text has been credited with the early transmission of much of Aristotle's natural philosophy—albeit an astrologized version—to the West. Rupescissa was greatly influenced by Abū Ma'shar's writings, particularly his account of celestial influence in *De magnis coniunctionibus*, which Rupescissa cites as the source of his own interpretation of great conjunctions and their apocalyptic effects on earth. See Richard Lemay, *Abu Ma'shar and Latin Aristotelianism in the Twelfth Century* (Beirut: American University, 1962), 59–60; on Abū Ma'shar's astrology, see John D. North, "Astrology and the Fortunes of Churches," *Centaurus* 24 (1980): 185–90; David Pingree, *The Thousands of Abū Ma'shar* (London: Warburg Institute, 1968). Abū Ma'shar's text has been edited and translated by Charles Burnett and Keiji Yamamoto as Abū Ma'shar, *On Historical Astrology: The Book of Religions and Dynasties (On the Great Conjunctions)*, 2 vols. (Leiden: Brill, 2000).

3. A number of illustrations of the "Zodiac Man" appear in fourteenth- and fifteenth-century medical manuals and treatises. On these images, see Charles Clark,

7. THE END OF NATURE

"The Zodiac Man in Medieval Medical Astrology," *Journal of the Rocky Mountain Medieval and Renaissance Association* 3 (1982): 13–38.

4. For general definitions of "*mirabilia*" and "*miracula*," see Caroline Walker Bynum, "Wonder," in *Metamorphosis and Identity* (New York: Zone Books, 2001), 41–50; Benedicta Ward, *Miracles and the Medieval Mind: Theory, Record, and Event, 1000–1215* (Philadelphia: University of Pennsylvania Press, 1982), 3–19. For views of nature as autonomous at Chartres and Saint-Victor, see M. D. Chenu, "Nature and Man: The Renaissance of the Twelfth Century," in *Nature, Man, and Society in the Twelfth Century*, trans. Jerome Taylor and Lester K. Little (Chicago: University of Chicago Press, 1968), 14–18. On the mendicants' approach to the study of nature, see Roger French and Andrew Cunningham, *Before Science: The Invention of the Friars' Natural Philosophy* (Aldershot: Scolar, 1996), esp. 70–98.

5. A discussion of the prophecy, along with its full text (drawn from the *Liber ostensor*) appears in Jean de Roquetaillade (John of Rupescissa), *Liber ostensor quod adesse festinant tempora*, ed. Clémence Thévenaz Modestin, Christine Morerod-Fattebert, André Vauchez, et al. (Rome: École Française de Rome, 2005) (cited hereafter as LO), 948–51.

6. See his account of the episode, LO, 141–43.

7. Jeanne Bignami-Odier, "Jean de Roquetaillade (de Rupescissa), Théologien, Polémiste, Alchimiste," *Histoire littéraire de la France* 41 (1981): 144. For more on the *Liber ostensor*'s political content, see the historical analyses that accompany the École Française de Rome edition of the *Liber ostensor*, esp. 33–44. See also Martin Aurell, "Prophétie et messianisme politique: La péninsule iberique au miroir du Liber ostensor de Jean de Roquetaillade," in *Textes prophétiques et la prophétie en Occident XIIe–XVIe siècle*, ed. André Vauchez (Rome: École Française de Rome, 1990), 27–71.

8. LO, 206: "Sequitur: *Tempestates, inundationes, sterilitas, distemperancia temporis, corruptio aeris ultra extremum poterunt videri.*"

9. On astrology in the Middle Ages, see John D. North, "Astrology and the Fortunes of Churches," 181–211; North, "Celestial Influence," 45–100; Laura Ackerman Smoller, *History, Prophecy, and the Stars: The Christian Astrology of Pierre d'Ailly, 1350–1420* (Princeton, N.J.: Princeton University Press, 1994); Tullio Gregory, "Temps astrologique et temps chrétien," in *Le temps chrétien de la fin de l'antiquité au moyen âge IIIe–XIIIe siècles*, ed. Jean-Marie Leroux (Paris: CNRS, 1984), 557–73; O. Neugebauer, *Astronomy and History: Selected Essays* (New York: Springer-Verlag, 1983); Anthony Grafton, "Starry Messengers: Recent Work in the History of Western Astrology," *Perspectives on Science* 8, no. 1 (2000): 70–83.

10. LO, 206–7. The planetary conjunction of 1345 was the subject of a number of astrological treatises. Several such writings (including those of John of Murs, John of Ashenden, and Levi ben Gerson) were addressed to Pope Clement VI in Avignon. Rupescissa was at the time confined in the papal prison and was aware of these texts, as the editors of the *Liber ostensor* have documented (LO, 209–11). We know from the *Liber ostensor* that Rupescissa was also familiar with Abū Ma'shar's *De magnis coniunctionibus*, linking him to the milieu of the papal astrologers of the 1350s. He also cites the *Liber horoscopi*

and an unidentified *libri judiciorum* on astronomy (LO, 210–12). On the *Liber horoscopus*, see Matthias Kaup, "Der *Liber Horoscopus*: Ein bildloser Übergang von der Diagrammatik zur Emblematik in der Tradition Joachims von Fiore," in *Die Bildwelt der Diagramme Joachims von Fiore: Zur Medialität religiös-politischer Programme im Mittelalter*, ed. Alexander Patschovsky (Ostfildern: Jan Thorbecke Verlag, 2003), 147–84. For conjunction theory in general, see Smoller, *History, Prophecy, and the Stars*, 20–22, 61–84. For Levi ben Gerson and his ties to Avignon, see Bernard R. Goldstein and David Pingree, eds., *Levi ben Gerson's Prognostication for the Conjunction of 1345, Transactions of the American Philosophical Society* vol. 80, no. 6 (Philadelphia: American Philosophical Society, 1990); Chris Schabel, "Philosophy and Theology Across Cultures: Gersonides and Auriol on Divine Foreknowledge," *Speculum* 81, no. 4 (2006): 1092–1117. For apocalypticism in the works of Abū Maʿshar and other Islamic astrologers, see Saïd Amir Arjomand, "Islamic Apocalypticism in the Classical Period," in *Apocalypticism in Western History and Culture*, ed. Bernard McGinn, vol. 2 of *The Encyclopedia of Apocalypticism* (New York: Continuum, 2000), 265–69.

11. This description follows very closely that of Smoller, *History, Prophecy, and the Stars*, 20–22; see also North, "Fortunes of Churches," 185–86.

12. Arjomand, "Islamic Apocalypticism," 265–71.

13. These symbolic conjunctions with Jupiter were not thought to be causative of the six religions, nor were they intended to refer to actual planetary conjunctions. They merely suggested which religions would arise. See Smoller, *History, Prophecy, and the Stars*, 61–62. For astrology and apocalypticism in general, also see Hilary M. Carey, "Astrology and Antichrist in the Later Middle Ages," in *Time and Eternity: The Medieval Discourse*, ed. G. Jaritz and G. Moreno-Riano (Turnhout: Brepols, 2003), 515–35; Carey, "Astrology and the Last Things in Later Medieval England and France," in *Prophecy, Apocalypse, and the Day of Doom: Proceedings of the 2000 Harlaxton Symposium*, ed. Nigel Morgan (Donington, U.K.: Watkins, 2004), 19–38.

14. See Roger Bacon, *The "Opus Majus" of Roger Bacon*, ed. John Henry Bridges, 3 vols. (Oxford: Clarendon Press, 1897) (hereafter cited as OM), 1:255–62; Bacon, *Opus tertium*, vol. 1 of *Opera quaedam hactenus inedita*, ed. J. S. Brewer (London: Longman, Green, Longman and Roberts, 1859), 272. See also Smoller, *History, Prophecy, and the Stars*, 61–62, and Mark T. Abate, "Roger Bacon and the Rage of Antichrist: The Apocalypse of a Thirteenth-Century Natural Philosopher," Ph.D. diss., Boston University, 1996, 112–13, 275–80.

15. OM, 1:268–69: "Nolo hic ponere os meum in coelum, sed scio quod si ecclesia vellet revolvere textum sacrum et prophetias sacras, atque prophetias Sibyllae, et Merlini et Aquilae, et Sestonis, Joachim et multorum aliorum, insuper historias et libros philosophorum, atque juberet considerari vias astronomiae, inveniretur sufficiens suspicio vel magis certitudo de tempore Antichristi."

16. Laura Ackerman Smoller, "The Alfonsine Tables and the End of the World: Astrology and Apocalyptic Calculation in the Later Middle Ages," in *The Devil, Heresy, and Witchcraft in the Middle Ages: Essays in Honor of Jeffrey B. Russell*, ed. Alberto Ferreiro (Leiden: Brill, 1998), 211–39.

17. Augustine of Hippo, *De civitate Dei*, trans. Henry Bettenson (Harmondsworth: Penguin, 1972), 2.137: "stellae significare potius ista quam facere, ut quasi locutio quaedam sit illa positio praedicens futura, non agens."

18. Roger Bacon argued that constellations "are not only signs, but do something in the way of excitation": according to Bacon's analysis, celestial species impressed form upon sublunary matter, which humans could counteract through free will and the power of the rational soul. For his discussion, see OM, 1:266–67. Bacon's assertion of celestial influence on the life of Christ may have provided the impetus for his censure between 1277 and 1279. On this, see Paul L. Sidelko, "The Condemnation of Roger Bacon," *Journal of Medieval History* 22 (1996): 69–81. See also Krzysztof Pomian, "Astrology as a Naturalistic Theology of History," in *'Astrologi hallucinati': Stars and the End of the World in Luther's Time*, ed. Paola Zambelli (Berlin: Walter de Gruyter, 1986), 32; David C. Lindberg, "Cosmology," in *Science in the Middle Ages*, ed. Lindberg (Chicago: University of Chicago Press, 1978), 288–90.

19. Paola Zambelli, introduction to *'Astrologi hallucinati': Stars and the End of the World in Luther's Time*, ed. Zambelli (Berlin: Walter de Gruyter, 1986), 22; Gregory, "Temps astrologique et temps chrétien," 564–65.

20. Smoller, "Of Earthquakes, Hail, Frogs, and Geography: Plague and the Investigation of the Apocalypse in the Later Middle Ages," in *Last Things: Death and the Apocalypse in the Middle Ages*, ed. Caroline Walker Bynum and Paul Freedman (Philadelphia: University of Pennsylvania Press, 2000), 165–66. For the context and consequences of the condemnations of 1277, see J. M. M. H. Thijssen, *Censure and Heresy at the University of Paris, 1200–1400* (Philadelphia: University of Pennsylvania Press, 1998), 40–56.

21. Josep Perarnau i Espelt, "El text primitiu del *De mysterio cymbalorum ecclesiae*, *Arxiu de Textos Catalans Antics* 7/8 (1988–89): 152–53: "Astrologi vero, qui probant quod motus retardationis octave spere compleri nequit in paucioribus annis quam in triginta sex millibus, debent scire quod suam potentiam et sapientiam Deus non alligavit naturalibus causis, sed sicut in productione mundi fuit supernaturaliter operatus, sic et in consummatione huius seculi supernaturaliter operabitur." Appendix 1 of Perarnau's article includes the full text of *De tempore adventus Antichristi*.

22. The criticism of Arnald by the masters of Paris and Oxford was fierce. Henry of Harclay was among the leaders of the scholastic reaction against Arnald at Oxford. His *quaestio*, *Utrum astrologi vel quicumque calculatores possint probare secundum adventum Christi*, accused Arnald of Jewish tendencies and even heresy. A fellow opponent of Arnald and a master at Paris, Peter du Croc of Avignon, also rejected all speculation about the time of the end and argued that Antichrist had already appeared in a mystical form. See Marjorie Reeves, *The Influence of Prophecy in the Later Middle Ages: A Study in Joachimism* (Oxford: Clarendon Press, 1969), 315–16.

23. Smoller, "Earthquakes," 166. Smoller has recently amended this observation after her discovery of a number of texts, including works by the astrologer John of Legnano, that predict Antichrist in the fourteenth century. See her "Astrology and the Sib-

yls: John of Legnano's *De adventu Christi* and the Natural Theology of the Later Middle Ages," *Science in Context* 20, no. 3 (2007): 423–50.

24. LO, 209–10: "dicunt philosophi quod conjunctio significavit infra XX annos tunc futuros novas sectas futures, altercationes horribiles de fidibus diversis. Et omnes sarraceni et judei et christiani astronomi, ex regulis per philosophos antiquos datis et in scriptis relictis, anno Domini MCCCXLV scribentes de influxu predicte conjunctionis, unanimiter nullo discrepante judicaverunt et in scriptis miserunt quod, ideo quia Saturnus per naturam significat Judeos et legem Judeorum—prout scribit Albumasar in tractatu primo *Libri conjuctionum et revolutionum annorum mundi*—, quia erat fortior Jove et Marte in sua domo et triplicitate, designavit quod infra XX annos qui finiunt anno Domini MCCCLXV exclusive debuit aperere conatus Judeorum et eorum falsus messias."

25. See John of Rupecissa, *Liber secretorum eventuum*, ed. Robert E. Lerner and Christine Morerod-Fattebert (Fribourg, Switzerland: Editions Universitaires, 1994), 87. On the eastern Antichrist in Rupescissa's writings, see chapter 2, pages 24–26; see also Louis Boisset, "Visions d'Orient chez Jean de Roquetaillade," in *Textes prophétiques et la prophétie en Occident XIIe-XVIe siècle*, ed. André Vauchez (Rome: École Française de Rome, 1990), 101–11. Rupescissa also claimed that some aspects of his own imprisonment had been predicted by an astrological conjunction in the sign of Aquarius: LO, 285–86.

26. LO, 207–8: "Et tamen expressus textus Scripture sacre est Job XLI capitulo quod ante aparitionem Antichristi, ista ipsum precedere debebent, ubi sub typo Leviatan, id est ballene marine, de Antichristo Dominus ita dicit: *Alitus ejus prunas ardere facit, et flamma de ore eius egreditur; in collo ejus morabitur fortitudo et faciem eius precedet egestas* (Job 41, 12–13). Si ergo *faciem ejus precedit egestas*, cum *egestas* proveniat per *tempestates* et *inundationes* et *distemperies temporis* et per *corruptiones aeris* et per *egestates mortalitates* proveniant, sequitur evidenter mira concordia inter naturam influxuum celorum et stellarum, ac oraculum istud et divinam Scripturam."

27. LO, 207: "Et ista ideo induxi ut clare ostendam miram concordiam inter naturam et divinam Scripturam pro predictis V tam pestiferis annis."

28. On Joachim scholarship, see chapter 2, note 16.

29. The other Joachite keystone is *plenitudo*, or fullness. See Bernard McGinn, *The Calabrian Abbot: Joachim of Fiore in the History of Western Thought* (New York: Macmillan, 1985), 22.

30. There are a number of studies of the Book of Nature in medieval literature. For an overview, see Constant Mews, "The World as Text: The Bible and the Book of Nature in Twelfth-Century Theology," in *Scripture and Pluralism: Reading the Bible in the Religiously Plural Worlds of the Middle Ages and Renaissance*, ed. Thomas J. Heffernan and Thomas E. Burman (Leiden: Brill, 2005), 95–122. Not all apocalypticists asserted the union of natural philosophy and divine truth as Rupescissa did. See, for instance, Peter John Olivi's anti-Aristotelianism in a spiritual context in O. Bettoni, "Olivi di fronte ad Aristotle," *Studi francescani* 55 (1958): 176–97.

7. THE END OF NATURE

31. Smoller, "Earthquakes," 163–66; R.W. Southern, "Aspects of the European Tradition of Historical Writing: 3. History as Prophecy," *Royal Historical Association Transactions* 5th ser., 22 (1972): 170–73.

32. Smoller, "Earthquakes," 177, 183–94.

33. Smoller, "Earthquakes," 184.

34. As quoted in Carey, "Astrology and the Last Things," 37.

35. DQE, 34: "Nec quinta Essentia, quam quaerimus, ad incorruptionem caeli omnino reducitur, sicut nec artificium adaequatur naturae: sed tamen incorruptibilis est respectu compositionis factae ex quatuor elementis."

36. DQE, 20: "Et dixi quod quintam Essentiam creavit Altissimus, quae extrahitur de corpore naturae creatae a Deo cum artificio humano."

37. Emphasis is mine. See chapter 6, page 123 for a fuller version of this quotation.

38. For a discussion of such claims, see William R. Newman's introduction to Pseudo-Geber, *The Summa Perfectionis of Pseudo-Geber: A Critical Edition, Translation and Study*, ed. William R. Newman (Leiden: Brill, 1991), 2–29; and Newman, *Promethean Ambitions: Alchemy and the Quest to Perfect Nature* (Chicago: University of Chicago Press, 2004), 34–114.

39. Constantine of Pisa, *The Book of the Secrets of Alchemy: Introduction, Critical Edition, Translation, and Commentary*, ed. Barbara Obrist (Leiden: Brill, 1990), 23–29. The edition of *Sciant artifices* prepared by E. J. Holmyard and D. C. Mandeville has been found to be inadequate; Newman provides a more reliable version in his Pseudo-Geber, *The Summa Perfectionis*, 48–51 (and an English translation, 2–3). See also Chiara Crisciani and Michela Pereira's discussion in the chapter "L'ingresso dell'alchimia in Occidente," in *L'arte del sole e della luna: alchimia e filosofia nel medioevo* (Spoleto: Centro italiano di studi sull'alto medioevo, 1996), 3–21.

40. Pseudo-Geber, *The Summa Perfectionis*, 2. An interpolation into the Latin translation of Avicenna's text adds that a metal could be transmuted if it was first reduced to prime matter and then changed into something new. This addition was seized upon by alchemists as a vindication of their efforts.

41. As quoted in William R. Newman, "The Homunculus and His Forebears: Wonders of Art and Nature," in *Natural Particulars: Nature and the Disciplines in Renaissance Europe*, ed. Anthony Grafton and Nancy Siraisi (Cambridge, Mass.: MIT Press, 1999), 323.

42. Pseudo-Geber, *The Summa Perfectionis*, 2–5; and Aristotle, *Physica*, 2.199a9–19. All citations to Aristotle can be found in *The Complete Works of Aristotle*, ed. Jonathan Barnes (Princeton, N.J.: Princeton University Press, 1984).

43. The revelation of Avicenna's authorship did not necessarily detract from the authority of the *Sciant artifices*; it continued to be included in arts curricula in the fourteenth century, as demonstrated by its appearance in the Oxford statutes of 1331. Obrist, in Constantine of Pisa, *The Book of the Secrets*, 26.

44. Pseudo-Geber, *The Summa Perfectionis*, 5.

45. Pseudo-Geber, *The Summa Perfectionis*, 21–29. This argument was expanded in the sixteenth century to cover the creation of artificial life forms, such as the basilisk and homunculus; see Newman, "Homunculus," 326–32.

7. THE END OF NATURE

46. Albertus Magnus, *Book of Minerals*, trans. Dorothy Wyckoff (Oxford: Clarendon Press, 1967), 178–79 (3.1.8). Albert also declared alchemy "the best imitator of nature" of all the arts (158 [3.1.2]), although nature is always "more certain and more direct than any art" (166 [3.1.5]).

47. For a more detailed account of this process, see Pearl Kibre, "Albertus Magnus on Alchemy," in *Albertus Magnus and the Sciences, Commemorative Essays 1980*, ed. James A. Weisheipl, O.P. (Toronto: Pontifical Institutes of Mediaeval Studies, 1980), 187–202; also Robert Halleux, "Albert le Grand et l'alchimie," *Revue des sciences philosophiques et théologiques* 66 (1982): 57–80.

48. For many alchemists, base metals were merely immature versions of gold, which was produced naturally in the bowels of the earth over thousands of years. The transmutation performed by the alchemist through heat was the same as the natural process of metal maturation, speeded up to accomplish the goal within a more convenient span of time. Albert the Great, however, denied that all metals had the same specific form and that gold was thus the mature version of all metals. He instead argued that in order to transmute metals, alchemists would have to corrupt the original form of the metal and then substitute a new form. Albert was, however, skeptical of transmutation. See William R. Newman, "Technology and Alchemical Debate in the Late Middle Ages," *Isis* 80 (1989): 431–32. See Albert's discussion in Albertus Magnus, *Book of Minerals*, 171–79 (3.1.7–9).

49. Petrus Bonus of Ferrara, *Pretiosa margarita novella*, in *Bibliotheca chemica curiosa*, ed. Jean-Jacques Manget (hereafter cited as BCC), 2 vols. (Geneva: Chouet et al., 1702), 2:19, 58.

50. Albertus Magnus, *Book of Minerals*, 178 (3.1.9); see also Newman's discussion in Pseudo-Geber, *The Summa Perfectionis*, 19.

51. Aristotle, *Meteorologica*, 4.381a24–b13. He writes that "broiling and boiling are artificial processes, but the same general kind of thing, as we said, is found in nature too. The affections produced are similar though they lack a name; for art imitates nature. For instance, the concoction of food in the body is like boiling, for it takes place in a hot and moist medium and the agent is the heat of the body."

52. John of Rupecissa, *Liber lucis*, in *Il libro della luce*, ed. Andrea Aromatico (Marsilio: Editori, 1997), 24v: "ut convertas omnia metalla imperfecta in melius argentum quam est illud quod est de minera."

53. Petrus Bonus of Ferrara, *Pretiosa margarita*, in BCC, 2:19.

54. Pseudo-Geber, *The Summa Perfectionis*, 30–35.

55. Pseudo-Geber, *The Summa Perfectionis*, 21; William R. Newman, "An Overview of Roger Bacon's Alchemy," in *Roger Bacon and the Sciences: Commemorative Essays*, ed. Jeremiah Hackett (Leiden: Brill, 1997), 318. For Bacon's apocalyptic and Joachite sympathies, see my discussion in chapter 6.

56. Quoted in Newman, "Homunculus," 336.

57. Pseudo-Geber, *The Summa Perfectionis*, 30–35.

58. DQE, 23: "Et aurum alchimicum, quod est ex corrosivis compositum, destruit naturam." See also DQE, 51 and 76.

7. THE END OF NATURE

59. DQE, 23.
60. Newman, "Technology," 434.
61. Chenu, "Nature and Man," 39–47.
62. DQE, 14: "Secretum primum est, quod per virtutem quam Deus contulit nature conditae, et humano magisterio subiectae, potest homo incommoda senectutis quibus nimis in antiquis viris Evangelicis impediuntur opera Evangelicae vitae curare, et amissam iuventutem iterum restaurare."
63. DQE, 28: "Et tunc cum Sole nostro et stellis terrenis facies operationem Dei recte, et in eius virtute miracula super terram, sicut inferius te docebo."
64. DQE, 120.
65. Gilbert de la Porrée made a parallel argument in his discussion of whether artificial products could be considered the works of God. He divided all works into three categories: creations of the divine, products of nature, and things made by men. All three were aspects of the divine governance, he concluded, and God was the sole creator. This sort of reasoning was also characteristic of the twelfth-century masters of Chartres, who similarly posited (following Plato in the *Timaeus*) that all things are the work of God but they can be divided into those things produced wholly by the divine, by the laws of nature, or by man imitating nature. See Chenu, "Nature and Man," 39–41.

8. CONCLUSION

1. For a history of early modern books of secrets, see William Eamon, *Science and the Secrets of Nature: Books of Secrets in Medieval and Early Modern Culture* (Princeton, N.J.: Princeton University Press, 1994).
2. See, for instance, Agostino Paravicini Bagliani, *The Pope's Body*, trans. David S. Peterson (Chicago: University of Chicago Press, 2000), 204–11; Pamela H. Smith, *The Business of Alchemy: Science and Culture in the Holy Roman Empire* (Princeton, N.J.: Princeton University Press, 1994).
3. On these manuscripts, see chapter 2, note 1.
4. It is interesting to note the similarities of Rupescissa's heirs and the twentieth-century millenarian groups studied by the sociologists Leon Festinger, Henry W. Riecken, and Stanley Schachter. According to their landmark study, *When Prophecy Fails* (Minneapolis: University of Minnesota Press, 1956), adherents to apocalyptic prophecy do not abandon their beliefs when the appointed date for the end passes without incident. Instead, they immediately find fault with their interpretation of a prophecy rather than with the prophecy itself. They usually reinterpret the prophecy to arrive at a future date for the eschaton. Strikingly, believers tend to become even more firm in their convictions and more eager to proselytize after such a failure.
5. Robert E. Lerner, *Feast of Saint Abraham: Medieval Millenarians and the Jews* (Philadelphia: University of Pennsylvania Press, 2000), 89–100.
6. Lerner, *Feast of Saint Abraham*, 91.

7. Lerner, *Feast of Saint Abraham*, 101–10.

8. Robert E. Lerner, "'Popular Justice': Rupescissa in Hussite Bohemia," in *Eschatologie und Hussitismus*, ed. Alexander Patschovsky, František Šmahel, and Antonín Hrubý (Prague: Historisches Institut, 1996), 42; and Lerner, "Medieval Millenarianism and Violence," in *Pace e guerra nel basso medioevo: Atti del XL convegno storico internazionale* (Spoleto: Centro italiano di studi sull'alto medioevo, 2004), 50.

9. On Telesphorus, see Emily Donckel, "Studien über die Prophezeiung des Fr. Telesforus von Cosenza, O.F.M. (1365–1386)," *Archivum Franciscanum Historicum* 26 (1933): 29–104, 282–314.

10. Gian Luca Potestà, "Radical Apocalyptic Movements in the Late Middle Ages," in *The Encyclopedia of Apocalypticism*, vol. 2 of *The Encyclopedia of Apocalypticism*, ed. Bernard McGinn (New York: Continuum, 2000), 124–26.

11. Lerner, "'Popular Justice,'" 43–49.

12. On this work, see Uwe Junker, ed., *Das "Buch der Heiligen Dreifaltigkeit" in seiner zweiten, alchemistischen Fassung (Kadolzburg 1433)* (Cologne: Institut für Geschichte der Medizin der Universität, 1986); Denis Duveen, "Le Livre de la très sainte trinité," *Ambix* 3 (1948): 26–32; Herwig Buntz, "Das Buch der heiligen Dreifaltigkeit: Sein Autor und seine Überlieferung," *Zeitschrift für deutsches Altertum und deutsche Litteratur* 101 (1972): 150–60; W. Ganzenmüller, "Das *Buch der heiligen Dreifaltigkeit*: Eine deutsche Alchemie aus dem Anfang des 15. Jahrhunderts," *Archiv für Kulturgeschichte* 29 (1939): 93–146. On the author and his connection to the Franciscan Spiritual legacy, see Barbara Obrist, *Les débuts de l'imagerie alchimique (XIVe-XVe siècles)* (Paris: Sycomore, 1982), 117–182, esp. 133.

13. Junker, *Das "Buch der Heiligen Dreifaltigkeit,"* 54, as cited in Michela Pereira, "Alchemy and the Use of Vernacular Languages in the Late Middle Ages," *Speculum* 74 (1999): 346.

14. For a printing history of *De quinta essentia* and the *Liber lucis*, see Robert Halleux, "Les ouvrages alchimiques de Jean de Roquetaillade," *Histoire littéraire de la France* 41 (1981): 242–44, 262–63, 268–76. Of the 142 copies of *De quinta essentia* listed by Robert Halleux, five date from the fourteenth century, two from the fourteenth to fifteenth, ninety-one from the fifteenth, four from the fifteenth to sixteenth, thirty-two from the sixteenth, and eight from the seventeenth. Of the fifty-five *Liber lucis* manuscripts, one copy dates from the fourteenth to fifteenth centuries, twenty-eight from the fifteenth century, twenty-one from the sixteenth, and five from the seventeenth and eighteenth.

15. The Gratarolo edition is the version upon which my readings are based. On Gratarolo, see Ian Maclean, "Heterodoxy in Natural Philosophy and Medicine: Pietro Pomponazzi, Guglielmo Gratarolo, Girolamo Cardano," in *Heterodoxy in Early Modern Science and Religion*, ed. John Brooke and Ian Maclean (Oxford: Oxford University Press, 2006), 17–19. The French version of *De quinta essentia* is now available as a modern reprint, *La vertu et propriété de la quinte essence de toutes choses* (Milan: Archè, 1971).

16. Halleux, "Ouvrages," 270–73.

17. As quoted in Michela Pereira, "*Medicina* in the Alchemical Writings Attributed to Raimond Lull (Fourteenth–Seventeenth Centuries)," in *Alchemy and Chemistry in the*

8. CONCLUSION

Sixteenth and Seventeenth Centuries, ed. Piyo Rattansi and Antonio Clericuzio (Dordrecht: Kluwer Academic Publishers, 1994), 1.

18. Danielle Jacquart, "Theory, Everyday Practice, and Three Fifteenth-Century Physicians," *Osiris* 2nd ser., 6 (1990): 153.

19. Paracelsus, *Archidoxis* (London: Thomas Brewster, 1660), 35.

20. Paracelsus, *Archidoxis*, 44.

21. Halleux, "Ouvrages," 275; Frank Sherwood Taylor, "The Idea of the Quintessence," in *Science, Medicine, and History: Essays on the Evolution of Scientific Thought and Medical Practice Written in Honour of Charles Singer*, ed. E. Ashworth Underwood (London: Oxford University Press, 1953), 263.

22. The English translation is quoted in Allen G. Debus, *The Chemical Philosophy: Paracelsian Science and Medicine in the Sixteenth and Seventeenth Centuries*, 2 vols. (New York: Science History Publications, 1977), 1:24.

23. Quoted in the chapter "Quintessence of Blood," in Piero Camporesi, *Juice of Life*, trans. Robert R. Barr (New York: Continuum, 1995), 29–31, 125nn. 4–6.

24. John Burnett, "The Giustiniani Medicine Chest," *Medical History* 26 (1982): 325–33.

25. Urszula Szulakowski, *The Sacrificial Body and the Day of Doom: Alchemy and Apocalyptic Discourse in the Protestant Reformation* (Leiden: Brill, 2006), 12, 17.

26. On Paracelsus and the Paracelsians, see Debus, *The Chemical Philosophy*; Debus, *Man and Nature in the Renaissance* (Cambridge: Cambridge University Press, 1978), 101–30; Walter Pagel, *Paracelsus: An Introduction to Philosophical Medicine in the Era of the Renaissance* (Basel: Karger, 1982); Gerhild Scholz Williams and Charles D. Gunnoe, eds., *Paracelsian Moments: Science, Medicine, and Astrology in Early Modern Europe* (Kirksville, Mo.: Truman State University Press, 2002).

27. Allen G. Debus, "Chemists, Physicians, and Changing Perspectives on the Scientific Revolution," *Isis* 89 (1998): 74.

28. See, for instance, John Noble Wilford, "Transforming the Alchemists: Serious Scientists Revisit a Discredited Craft," *New York Times*, August 1, 2006.

29. Despite the Paracelsians' renunciation of previous authorities, scholars such as Allen Debus have demonstrated their unacknowledged debt to the ancients. See *Chemical Philosophy*, 1:51–204.

BIBLIOGRAPHY

PRIMARY SOURCES

Manuscripts

John of Rupescissa. *Liber ostensor*. Vatican City, Biblioteca Apostolica Vaticana, Rossiano MS 753, fol. 1–149.
——. *Vade mecum in tribulatione*. Tours, Bibliothèque municipale, MS 520, fol. 32v–47v.
——. *Vade mecum in tribulatione*. Vatican City, Biblioteca Apostolica Vaticana, Reg. Lat. 1964, fol. 196–203v.
——. *Vade mecum in tribulatione*. Vatican City, Biblioteca Apostolica Vaticana, Lat. 4265, fol. 175–181v.
——. *Vade mecum in tribulatione*. Venice, Biblioteca Nazionale Marciana, MS Lat. III, 39, fol. 1–19v.

Printed Sources

Abū Ma'shar. *On Historical Astrology: The Book of Religions and Dynasties (On the Great Conjunctions)*. Ed. and trans. Charles Burnett and Keiji Yamamoto. 2 vols. Leiden: Brill, 2000.
Albertus Magnus. *Book of Minerals*. Trans. Dorothy Wyckoff. Oxford: Clarendon Press, 1967.
Angelo of Clareno. *Epistole*. Vol. 1 of *Angeli Clareni Opera*, ed. Lydia Von Auw. Rome: Nella sede dell'Istituto, 1980.
Arano, Luisa Cogliati, ed. *The Medieval Health Handbook: Tacuinum sanitatis*. New York: George Braziller, 1976.
Aristotle. *The Complete Works of Aristotle*. Ed. Jonathan Barnes. Princeton, N.J.: Princeton University Press, 1984.

Arnald of Vilanova. "*L'Allocutio christini* d'Arnau de Vilanova: edició i estudi del text." Ed. Josep Perarnau i Espelt. *Arxiu de Textos Catalans Antics* 11 (1992): 7–135.

——. *Aphorismi de gradibus.* Ed. Michael R. McVaugh et al. Granada: Seminarium Historiae Medicae Granatensis, 1975.

——. *Arnaldi de Villanova Opera medica omnia.* Ed. Michael R. McVaugh et al. Granada: Seminarium Historiae Medicae Granatensis, 1981–.

——. *Epistola de dosi tyriacalium medicinarum.* In Vol. 3 of *Arnaldi de Villanova Opera medica omnia*, ed. Michael R. McVaugh, 55–91. Barcelona: Seminarium Historiae Medicae Cantabricensis, 1985.

(Pseudo-)Arnald of Vilanova. *The Earliest Printed Book on Wine.* Ed. Henry E. Sigerist. New York: Schuman's, 1943.

——. *Epistola Magistri Arnaldi Cathelani de Villanova ad Magistrum Iacobum de Toleto de sanguine humano.* Basel, 1561.

——. *Le Rosier alchimique de Montpellier (Lo Rosari): Textes, traduction, notes, et commentaires.* Ed. Antoine Calvet. Paris: Presses de l'Université de Paris-Sorbonne, 1997.

——. *Tractatus de vinis.* Paris: Felix Baligault pour Claude Jaumar et Thomas Julian, 1500.

Augustine of Hippo. *De civitate Dei.* Trans. Henry Bettenson. Harmondsworth: Penguin, 1972.

Bacon, Roger. *Fratris Rogeri Bacon De retardatione accidentium senectutis cum aliis opusculis de rebus medicinalibus.* Ed. A. G. Little and E. Withington. Oxford: Clarendon Press, 1928.

——. *In libro sex scientarum in 3° gradu sapientie.* In *Fratris Rogeri Bacon De retardatione accidentium senectutis cum aliis opusculis de rebus medicinalibus*, ed. A. G. Little and E. Withington, 181–86. Oxford: Clarendon Press, 1928.

——. *The "Opus Majus" of Roger Bacon.* Ed. John Henry Bridges. 3 vols. Oxford: Clarendon Press, 1897.

——. *Opus minus.* Vol. 2 of *Opera quaedam hactenus inedita Rogeri Baconi*, ed. J. S. Brewer. London: Longman, Green, Longman, and Roberts, 1859.

——. *Opus tertium.* Vol. 1 of *Opera quaedam hactenus inedita Rogeri Baconi*, ed. J. S. Brewer. London: Longman, Green, Longman, and Roberts, 1859.

——. *Part of the* Opus tertium *of Roger Bacon, Including a Fragment Now Printed for the First Time.* Ed. A. G. Little. Aberdeen: University Press, 1912.

——. *Roger Bacon's Philosophy of Nature: A Critical Edition, with English Translation, Introduction, and Notes of* De multiplicatione specierum *and* De speculis comburentibus. Ed. David C. Lindberg. Oxford: Clarendon Press, 1983.

——. *Secretum secretorum cum glossis et notulis.* Vol. 5 of *Opera hactenus inedita Rogeri Baconi*, ed. Robert Steele. Oxford: Clarendon Press, 1909.

Constantine of Pisa. *The Book of the Secrets of Alchemy: Introduction, Critical Edition, Translation, and Commentary.* Ed. Barbara Obrist. Leiden: Brill, 1990.

Goldstein, Bernard R., and David Pingree, eds. *Levi ben Gerson's* Prognostication for the Conjunction of 1345, *Transactions of the American Philosophical Society*, vol. 80, no. 6. Philadelphia: American Philosophical Society, 1990.

Gui, Bernard. *Manuel de l'inquisiteur.* Ed. G. Mollat. 2 vols. Paris: Les Belles Lettres, 1877–1968.

Guardo, Alberto Alonso. *Los pronósticos médicos en la medicina medieval: El Tractatus de crisi et de diebus creticis de Bernardo de Gordonio.* Lingüística y Filología 54. Valladolid: Secretariado de Publicaciones e Intercambio Editorial, Universidad de Valladolid, 2003.

Hildegard of Bingen. *Hildegardis Bingensis Liber divinorum operum.* Ed. A. Derolez and P. Dronke. Corpus Christianorum. Continuatio Medievalis, vol. 92. Turnhout: Brepols, 1996.

Horrox, Rosemary, ed. *The Black Death.* Manchester: Manchester University Press, 1994.

Israeli, Isaac. *Omnia opera Ysaac,* 2 vols. Leiden, 1515.

Jean de Venette. *The Chronicle of Jean de Venette.* Ed. Richard A. Newhall. Trans. Jean Birdsall. New York: Columbia University Press, 1953.

John of Rupescissa. *Johannes' de Rupescissa "Liber de consideratione quintae essentiae omnium rerum" deutsch: Studien zur Alchemia medica des 15. Bis 17. Jahrhunderts mit kritische Edition des Textes.* Ed. Udo Benzenhöfer. Stuttgart: Franz Steiner, 1989.

——. *Liber de consideratione quintae essentiae omnium rerum.* Basel, 1561.

——. *Liber lucis.* In *Il libro della luce,* ed. Andrea Aromatico. Marsilio: Editori, 1997.

——. *Liber ostensor quod adesse festinant tempora.* Ed. Clémence Thévenaz Modestin, Christine Morerod-Fattebert, André Vauchez, et al. Rome: École Française de Rome, 2005.

——. *Liber secretorum eventuum.* Ed. Robert E. Lerner and Christine Morerod-Fattebert. Fribourg: Editions Universitaires, 1994.

——. *Reverendissime pater.* In *Appendix ad fasciculum rerum expetendarum et fugiendarum,* ed. Edward Brown, 2:495–96. London: Richard Chiswell, 1690.

——. *Vade mecum in tribulationem.* In *Appendix ad fasciculum rerum expetendarum et fugiendarum,* ed. Edward Brown, 2:496–508. London: Richard Chiswell, 1690.

——. *La vertu et propriété de la quinte essence de toutes choses.* Milan: Archè, 1971.

——. *Vos misistis.* In *Appendix ad fasciculum rerum expetendarum et fugiendarum,* ed. Edward Brown, 2:494. London: Richard Chiswell, 1690.

Junker, Uwe, ed. *Das "Buch der Heiligen Dreifaltigkeit" in seiner zweiten, alchemistischen Fassung (Kadolzburg 1433).* Cologne: Institut für Geschichte der Medizin der Universität, 1986.

Llull, Ramón. *Il "Testamentum" alchemico attribuito a Raimondo Lullo: Edizione del testo latino e catalano dal manoscritto Oxford, Corpus Christi College, 244.* Ed. Michela Pereira and Barbara Spaggiari. Florence: Sismel, 1999.

——. *Selected Works.* Ed. and trans. Anthony Bonner. Princeton, N.J.: Princeton University Press, 1985.

(Pseudo-)Llull, Ramón. *Raimundi Lulii Maiorici, philosophi acutissimi, medicique celeberrimi, De secretis naturae siue quinta essentia libri duo.* Venice: Petrus Schoeffer, 1542.

Manget, Jean-Jacques, ed. *Bibliotheca chemica curiosa.* 2 vols. Geneva: Chouet et al., 1702.

McGinn, Bernard, ed. *Visions of the End: Apocalyptic Traditions in the Middle Ages.* New York: Columbia University Press, 1998.

O'Boyle, Cornelius. *Medieval Prognosis and Astrology: A Working Edition of Aggregationes de crisi et creticis diebus, With Introduction and English Summary.* Cambridge: Wellcome Unit for the History of Medicine, 1991.

Paracelsus. *Archidoxis.* London: Thomas Brewster, 1660.

Paré, Ambroise. *Oeuvres complètes d'Ambroise Paré.* Ed. J.-F. Malgaigne. 3 vols. Paris: J.-B. Baillière, 1840–1841.

Petroff, Elizabeth, ed. *Medieval Women's Visionary Literature.* New York: Oxford University Press, 1986.

Petrus Bonus of Ferrara. *Pretiosa margarita novella.* In *Bibliotheca chemica curiosa,* ed. Jean-Jacques Manget, 2:1–79. Geneva: Chouet et al., 1702.

——. *Preziosa Margarita Novella: Edizione del volgarizzamento, introduzione e note.* Ed. Chiara Crisciani. Florence: Nuova Italia Editrice, 1976.

Pseudo-Geber. *The Summa Perfectionis of Pseudo-Geber: A Critical Edition, Translation, and Study.* Ed. William R. Newman. Leiden: Brill, 1991.

Pseudo-Methodius, *Die Apokalypse des Pseudo-Methodius, die ältesten griechischen und lateinischen Übersetzungen.* Ed. W. J. Aerts and G. A. A. Kortekaas. Louvain: Peeters, 1998.

SECONDARY SOURCES

Abate, Mark T. "Roger Bacon and the Rage of Antichrist: The Apocalypse of a Thirteenth-Century Natural Philosopher." Ph.D. diss., Boston University, 1996.

Aberth, John. *The Black Death: The Great Mortality of 1348–1350, a Brief History with Documents.* New York: Palgrave, 2005.

Alexander, Paul. "Medieval Apocalypses as Historical Sources." *American Historical Review* 73 (1968): 997–1018.

Allmand, C. T. *The Hundred Years War: England and France at War, c. 1300–c. 1450.* Cambridge: Cambridge University Press, 1988.

Arjomand, Saïd Amir. "Islamic Apocalypticism in the Classical Period." In *Apocalypticism in Western History and Culture,* vol. 2 of *The Encyclopedia of Apocalypticism,* ed. Bernard McGinn, 238–83. New York: Continuum, 2000.

Aromatico, Andrea. "Premessa al testo." In *Il libro della luce,* ed. Andrea Aromatico, 116–19. Marsilio: Editori, 1997.

Arrizabalaga, Jon. "Facing the Black Death: Perceptions and Reactions of University Medical Practitioners." In *Practical Medicine from Salerno to the Black Death,* ed. Luis García-Ballester et al., 241–53. Cambridge: Cambridge University Press, 1994.

Aurell, Martin. "Prophétie et messianisme politique: La péninsule iberique au miroir du Liber ostensor de Jean de Roquetaillade." In *Textes prophétiques et la prophétie en Occident XIIe–XVIe siècle,* ed. André Vauchez, 27–71. Rome: École Française de Rome, 1990.

Backman, Clifford. "Arnau de Vilanova and the Franciscan Spirituals in Sicily." *Franciscan Studies* 50 (1990): 3–29.
———. *The Decline and Fall of Medieval Sicily: Politics, Religion, and Economy in the Reign of Frederick III, 1296–1337*. Cambridge: Cambridge University Press, 1995.
———. "The Reception of Arnau de Vilanova's Religious Ideas." In *Christendom and Its Discontents: Exclusion, Persecution, and Rebellion, 1000–1500*, ed. Scott L. Waugh and Peter D. Diehl, 112–31. Cambridge: Cambridge University Press, 1995.
Badía, Lola, and Anthony Bonner. *Ramón Llull: Vida, pensamiento y obra literaria*. Barcelona: Sirmio, 1992.
Baldwin, John. *Scholastic Culture of the Middle Ages, 1000–1300*. Lexington, Mass.: Heath, 1971.
Barnay, Sylvie. "L'univers visionnaire de Jean de Roquetaillade." In *Fin du monde et signes des temps: Visionnaires et prophètes en France méridionale fin XIIIe–début XVe siècle, Cahiers de Fanjeaux* 27 (1992): 171–90.
Bettoni, O. "Olivi di fronte ad Aristotle." *Studi francescani* 55 (1958): 176–97.
Bignami-Odier, Jeanne. *Études sur Jean de Roquetaillade (Johannes de Rupescissa)*. Paris: Vrin, 1952.
———. "Jean de Roquetaillade (de Rupescissa), Théologien, Polémiste, Alchimiste." *Histoire littéraire de la France* 41 (1981): 75–240.
Binski, Paul. *Medieval Death: Ritual and Representation*. Ithaca, N.Y.: Cornell University Press, 1996.
Biraben, Jean Noël. *Les hommes et la peste en France et dans les pays européens et méditerranéens*. 2 vols. Paris: Mouton, 1975–1976.
Boilloux, Marc. "Étude d'une commentaire prophétique de XIV siècle: Jean de Roquetaillade et l'*Oracle Cyrille* (v. 1345–1349)." Ph.D. diss., École des Chartres, 1993.
Boisset, Louis. "Visions d'Orient chez Jean de Roquetaillade." In *Textes prophétiques et la prophétie en Occident XIIe–XVIe siècle*, ed. André Vauchez, 101–11. Rome: École Française de Rome, 1990.
Bollweg, John August. "Sense of a Mission: Arnau of Vilanova on the Conversion of Muslims and Jews." In *Iberia and the Mediterranean World of the Middle Ages: Studies in Honor of Robert I. Burns, S.J.*, ed. Larry J. Simon, 50–71. Leiden: Brill, 1995.
Bono, James J. *The Word of God and the Languages of Man: Interpreting Nature in Early Modern Science and Medicine*. Vol. 1: *Ficino to Descartes*. Madison: University of Wisconsin Press, 1995.
Boudet, J.-P. "La papauté d'Avignon et l'astrologie." In *Fin du monde et signes des temps: Visionnaires et prophètes en France méridionale fin XIIIe–début XVe siècle, Cahiers de Fanjeaux* 27 (1992): 257–93.
Brann, Noel L. "Alchemy and Melancholy in Mediaeval and Renaissance Thought: A Query Into the Mystical Basis of their Relationship." *Ambix* 32 (1985): 127–48.
Brehm, E. "Roger Bacon's Place in the History of Alchemy." *Ambix* 23 (1976): 53–58.
Brooke, Rosalind B. *The Coming of the Friars*. London: George Allen and Unwin, 1975.

Brown, Elizabeth A. R. "Death and the Human Body in the Later Middle Ages: The Legislation of Boniface VIII on the Division of the Corpse." *Viator* 12 (1981): 221–70.
Brown, Peter. *The Body and Society: Men, Women, and Sexual Renunciation in Early Christianity.* New York: Columbia University Press, 1988.
Brown, Theodore L. *Making Truth: Metaphor in Science.* Urbana: University of Illinois Press, 2003.
Buntz, Herwig. "*Das Buch der heiligen Dreifaltigkeit*: Sein Autor und seine Überlieferung." *Zeitschrift für deutsches Altertum und deutsche Litteratur* 101 (1972): 150–60.
Burnett, John. "The Giustiniani Medicine Chest." *Medical History* 26 (1982): 325–33.
Burnham, Louisa A. *So Great a Light, So Great a Smoke: The Beguin Heretics of Languedoc.* Ithaca, N.Y.: Cornell University Press, 2008.
Burr, David. "Bonaventure, Olivi, and Franciscan Eschatology." *Collectanea Franciscana* 53 (1983): 23–40.
———. *Olivi and Franciscan Poverty: The Origins of the Usus Pauper Controversy.* Philadelphia: University of Pennsylvania Press, 1989.
———. "Olivi, Apocalyptic Expectation, and Visionary Experience." *Traditio* 41 (1985): 273–88.
———. "Olivi's Apocalyptic Timetable." *Journal of Medieval and Renaissance Studies* 11 (1981): 237–60.
———. *Olivi's Peaceable Kingdom: A Reading of the Apocalypse Commentary.* Philadelphia: University of Pennsylvania Press, 1993.
———. *The Persecution of Peter Olivi*, Transactions of the American Philosophical Society, vol. 66, no. 5. Philadelphia: American Philosophical Society, 1976.
———. *The Spiritual Franciscans.* University Park: Pennsylvania State University Press, 2001.
Bynum, Caroline Walker. "The Blood of Christ in the Middle Ages." *Church History* 71 (2002): 685–714.
———. *Fragmentation and Redemption: Essays on Gender and the Human Body in Medieval Religion.* New York: Zone Books, 1992.
———. *Holy Feast and Holy Fast: The Religious Significance of Food to Medieval Women.* Berkeley: University of California Press, 1987.
———. "Metamorphosis, or Gerald and the Werewolf." In *Metamorphosis and Identity,* 77–111. New York: Zone Books, 2001.
———. *The Resurrection of the Body in Western Christianity, 200–1336.* New York: Columbia University Press, 1995.
———. "Violent Imagery in Late Medieval Piety." *German Historical Institute Bulletin* 30 (2002): 1–34.
———. "Wonder." In *Metamorphosis and Identity,* 37–75. New York: Zone Books, 2001.
———. *Wonderful Blood: Theology and Practice in Late Medieval Northern Germany and Beyond.* Philadelphia: University of Pennsylvania Press, 2007.
Bynum, Caroline Walker, and Paul Freedman. *Last Things: Death and the Apocalypse in the Middle Ages.* Philadelphia: University of Pennsylvania Press, 2000.

Calvet, Antoine. "L'alchimie d'Arnaud de Villeneuve." In *Terres médiévales*, ed. Ribémont Bernard, 21–33. Paris: Klincksieck, 1993.

———. "Alchimie et joachimisme dans les alchimica pseudo-arnaldiens." In *Alchimie et philosophie à la Renaissance*, ed. Jean-Claude Margolin and Sylvain Matton, 93–107. Paris: Vrin, 1993.

———. "Mutations de l'alchimie medicale au XVe siècle: a propos des textes authentiques et apocryphes d'Arnaud de Villeneuve." *Micrologus* 3 (1995): 185–209.

Camporesi, Piero. *Juice of Life: The Symbolic and Magic Significance of Blood*. Trans. Robert R. Barr. New York: Continuum, 1995.

Carey, Hilary M. "Astrology and Antichrist in the Later Middle Ages." In *Time and Eternity: The Medieval Discourse*, ed. G. Jaritz and G. Moreno-Riano, 515–35. Turnhout: Brepols, 2003.

———. "Astrology and the Last Things in Later Medieval England and France." In *Prophecy, Apocalypse, and the Day of Doom: Proceedings of the 2000 Harlaxton Symposium*, ed. Nigel Morgan, 19–38. Donington, U.K.: Watkins, 2004.

Casteen, Elizabeth. "John of Rupescissa's Letter *Reverendissime pater* (1350) in the Aftermath of the Black Death." *Franciscana* 6 (2004): 139–84.

Cazenave, Annie. "La vision eschatologique des spirituels franciscains autour de leur condamnation." In *The Use and Abuse of Eschatology in the Middle Ages*, ed. Werner Verbeke, Daniel Verhelst, and Andries Welkenhuysen, 393–403. Leuven: Leuven University Press, 1988.

Chenu, M. D. "Nature and Man: The Renaissance of the Twelfth Century." In *Nature, Man, and Society in the Twelfth Century*, trans. Jerome Taylor and Lester K. Little, 1–48. Chicago: University of Chicago Press, 1968.

Clark, Charles. "The Zodiac Man in Medieval Medical Astrology." *Journal of the Rocky Mountain Medieval and Renaissance Association* 3 (1982): 13–38.

Cohn, Samuel K. *The Black Death Transformed: Disease and Culture in Early Renaissance Europe*. New York: Oxford University Press, 2002.

———. *Creating the Florentine State: Peasants and Rebellion, 1348–1434*. Cambridge: Cambridge University Press, 1999.

———. *Lust for Liberty: The Politics of Social Revolt in Medieval Europe, 1200–1425: Italy, France, and Flanders*. Cambridge, Mass.: Harvard University Press, 2006.

Collins, Adela Yarbro. "The Early Christian Apocalypses." *Semeia* 14 (1979): 61–121.

Collins, John J. *The Apocalyptic Imagination: An Introduction to Jewish Apocalyptic Literature*. 1984. Reprint, Grand Rapids, Mich.: Eerdmans, 1998.

———. "Introduction: Towards the Morphology of a Genre." *Semeia* 14 (1979): 1–20.

Constable, Giles. *Three Studies in Medieval Religious and Social Thought*. Cambridge: Cambridge University Press, 1995.

Courtenay, William J. "Curers of Body and Soul: Medical Doctors as Theologians." In *Religion and Medicine in the Middle Ages*, ed. Peter Biller and Joseph Ziegler, 69–75. York: York Medieval Press, 2001.

Crisciani, Chiara. "Alchemy and Medicine in the Middle Ages: Recent Studies and Projects for Research." *Bulletin de philosophie médiévale* 38 (1996): 9–21.

———. "The Conception of Alchemy as Expressed in the *Pretiosa Margarita Novella* of Petrus Bonus of Ferrara." *Ambix* 20 (1973): 165–81.

———. "From the Laboratory to the Library: Alchemy According to Guglielmo Fabri." In *Natural Particulars: Nature and the Disciplines in Renaissance Europe*, ed. Anthony Grafton and Nancy Siraisi, 295–319. Cambridge, Mass.: MIT Press, 1999.

———. "History, Novelty, and Progress in Scholastic Medicine." *Osiris* 2nd ser., 6 (1990): 118–39.

———. *Il Papa e l'alchimia: Felice V, Guglielmo Fabri e l'elixir*. Rome: Viella, 2002.

———. "La *quaestio de alchimia* fra Duecento e Trecento." *Medioevo* 2 (1976): 119–68.

Crisciani, Chiara, and Michela Pereira. *L'arte del sole e della luna: alchimia e filosofia nel medioevo*. Spoleto: Centro italiano di studi sull'alto medioevo, 1996.

———. "Black Death and Golden Remedies: Some Remarks on Alchemy and the Plague." In *The Regulation of Evil: Social and Cultural Attitudes to Epidemics in the Late Middle Ages*, ed. Agostino Paravicini Bagliani and Francesco Santi, 7–39. Florence: Sismel, 1998.

Daniel, E. Randolph. *The Franciscan Concept of Mission in the High Middle Ages*. Lexington: University Press of Kentucky, 1975.

Debus, Allen G. *The Chemical Philosophy: Paracelsian Science and Medicine in the Sixteenth and Seventeenth Centuries*. 2 vols. New York: Science History Publications, 1977.

———. "Chemists, Physicians, and Changing Perspectives on the Scientific Revolution." *Isis* 89 (1998): 66–81.

———. *Man and Nature in the Renaissance*. Cambridge: Cambridge University Press, 1978.

Demaitre, Luke. "The Care and Extension of Old Age in Medieval Medicine." In *Aging and the Aged in Medieval Europe*, ed. Michael M. Sheehan, 3–22. Toronto: Pontifical Institute of Mediaeval Studies, 1990.

———. *Doctor Bernard de Gordon: Professor and Practitioner*. Toronto: Pontifical Institute of Mediaeval Studies, 1980.

Dobbs, Betty Jo Teeter. *Alchemical Death and Resurrection: The Significance of Alchemy in the Age of Newton*. Washington, D.C: Smithsonian Institution Libraries, 1990.

———. *The Foundations of Newton's Alchemy; Or, "The Hunting of the Greene Lyon."* Cambridge: Cambridge University Press, 1975.

———. *The Janus Faces of Genius: The Role of Alchemy in Newton's Thought*. Cambridge: Cambridge University Press, 1991.

Donckel, Emily. "Studien über die Prophezeiung des Fr. Telesforus von Cosenza, O.F.M. (1365–1386)." *Archivum Franciscanum Historicum* 26 (1933): 29–104, 282–314.

Duveen, Denis. "Le Livre de la très sainte trinité." *Ambix* 3 (1948): 26–32.

Eamon, William. *Science and the Secrets of Nature: Books of Secrets in Medieval and Early Modern Culture*. Princeton, N.J.: Princeton University Press, 1994.

Easton, Stewart C. *Roger Bacon and His Search for a Universal Science*. New York: Columbia University Press, 1952.

Emmerson, Richard K. *Antichrist in the Middle Ages: A Study of Medieval Apocalypticism, Art, and Literature.* Seattle: University of Washington Press, 1981.

Esser, Cajetan, O.F.M. *Origins of the Franciscan Order.* Trans. Aedan Daly and Irina Lynch. Chicago: Franciscan Herald Press, 1970.

Festinger, Leon, Henry W. Riecken, and Stanley Schachter. *When Prophecy Fails.* Minneapolis: University of Minnesota Press, 1956.

Franciscains d'Oc: Les spirituels ca. 1280–1324, Cahiers de Fanjeaux 10 (1975).

French, Roger. "Astrology in Medical Practice." In *Practical Medicine from Salerno to the Black Death,* ed. Luis García-Ballester et al., 30–59. Cambridge: Cambridge University Press, 1994.

French, Roger, and Andrew Cunningham. *Before Science: The Invention of the Friars' Natural Philosophy.* Aldershot: Scolar, 1996.

Fried, Johannes. *Aufstieg aus dem Untergang: apokalyptisches Denken und die Entstehung der modernen Naturwissenschaft im Mittelalter.* Munich: C. H. Beck, 2001.

Gagnon, Claude. "Alchimie, techniques, et technologie." In *Les arts mécaniques au moyen âge.* Cahiers d'études médiévales, vol. 7:131–46. Paris: Vrin, 1982.

Ganzenmüller, W. *Die Alchemie im Mittelalter.* Paderborn: Verlag der Bonifaciusdruckerei, 1938.

———. *L'Alchimie au moyen âge.* Trans. G. Petit-Dutaillis. Paris: Aubier, 1940.

———. "Das *Buch der heiligen Dreifaltigkeit*: Eine deutsche Alchemie aus dem Anfang des 15. Jahrhunderts." *Archiv für Kulturgeschichte* 29 (1939): 93–146.

García-Ballester, Luis. "On the Origin of the 'Six Non-natural Things' in Galen." In *Galen and Galenism: Theory and Practice from Antiquity to the European Renaissance,* ed. Jon Arrizabalaga et al., 105–15. Burlington, Vt.: Variorum, 2002.

Gardiner, Eileen ed. *Visions of Heaven and Hell Before Dante.* New York: Italica Press, 1989.

Gentner, Dedre. and Michael Jeziorski. "The Shift from Metaphor to Analogy in Western Science." In *Metaphor and Thought,* ed. Andrew Ortony, 447–80. Cambridge: Cambridge University Press, 1993.

Getz, Faye Marie. "The Faculty of Medicine Before 1500." In *Late Medieval Oxford,* vol. 2. of *History of the University of Oxford,* ed. J. I. Catto and Ralph Evans, 373–406. New York: Clarendon Press, 1992.

———. "To Prolong Life and Promote Health: Baconian Alchemy and the English Learned Tradition." In *Health, Disease, and Healing in Medieval Culture,* ed. S. Campbell, 135–45. New York: St. Martin's Press, 1992.

Giralt Soler, Sebastià. "Un alquimista medieval per als temps moderns: Les edicions del corpus alquímic atribuït a Arnau de Vilanova en llur context (c. 1477–1754)." *Arxiu de textos catalans antics* 23/24 (2005): 61–128.

Gougaud, Louis. *Devotional and Ascetic Practices in the Middle Ages.* Trans. G. C. Bateman London: Burns, Oates, and Washbourne, 1927.

Grafton, Anthony. "Starry Messengers: Recent Work in the History of Western Astrology." *Perspectives on Science* 8, no. 1 (2000): 70–83.

Grafton, Anthony, and Nancy Siraisi, eds. *Natural Particulars: Nature and the Disciplines in Renaissance Europe*. Cambridge, Mass.: MIT Press, 1999.

Grant, Edward. *The Foundations of Modern Science in the Middle Ages: Their Religious, Institutional, and Intellectual Contexts*. Cambridge: Cambridge University Press, 1996.

Gregory, Tullio. *Anima mundi: La filosofia di Guglielmo di Conches e la scuola di Chartres*. Florence: G. C. Sansoni, 1955.

———. "Temps astrologique et temps chrétien." In *Le temps chrétien de la fin de l'antiquité au moyen âge IIIe–XIIIe siècles*, ed. Jean-Marie Leroux, 557–73. Paris: CNRS, 1984.

Gruman, Gerald J. *A History of Ideas About the Prolongation of Life: The Evolution of Prolongevity Hypotheses to 1800*, Transactions of the American Philosophical Society, vol. 56, no. 9. Philadelphia: American Philosophical Society, 1966.

Gwei-Djen, Lu, Joseph Needham, and Dorothy Needham. "The Coming of Ardent Water." *Isis* 19 (1972): 69–112.

Hackett, Jeremiah, ed. *Roger Bacon and the Sciences: Commemorative Essays*. Leiden: Brill, 1997.

———. "Roger Bacon on *Scientia Experimentalis*." In *Roger Bacon and the Sciences: Commemorative Essays*, ed. Hackett, 277–315. Leiden: Brill, 1997.

Hall, Thomas. "Life, Death, and the Radical Moisture." *Clio medica* 6 (1971): 3–23.

Halleux, Robert. "Albert le Grand et l'alchimie." *Revue des sciences philosophiques et théologiques* 66 (1982): 57–80.

———. "Les ouvrages alchimiques de Jean de Roquetaillade." *Histoire littéraire de la France* 41 (1981): 241–84.

———. *Papyrus de Leyde, Papyrus de Stockholm: Fragments de recettes*. Vol. 1 of *Les alchimistes grecs*. Paris: Les Belles Letters, 1981.

———. "Problèmes de lexicographie alchimiste." In *La lexicographie du latin médiéval et ses rapports avec les recherches actuelles sur la civilisation du Moyen-Age*, ed. Yves Lefèvre, 355–65. Paris: CNRS, 1981.

———. *Les textes alchimiques*. Typologie des sources du moyen âge occidental, vol. 32. Turnhout: Brepols, 1979.

Halversen, Marguerite Ann. "'The Consideration of Quintessence': An Edition of a Middle English Translation of John of Rupescissa's *Liber de Consideratione Quintae Essentiae Omnium Rerum* with Introduction, Notes, and Commentary." Ph.D. diss., Michigan State University, 1998.

Hamburger, Jeffrey F. *Nuns as Artists: The Visual Culture of a Medieval Convent*. Berkeley: University of California Press, 1997.

Harvey, E. Ruth. *The Inward Wits: Psychological Theory in the Middle Ages and the Renaissance*. London: Warburg Institute, 1975.

Haureau, Barthélemy. "Arnauld de Villeneuve, médecin et chimiste." *Histoire littéraire de la France* 28 (1881): 26–126.

Hennessy, Marlene Villalobos. "Aspects of Blood Piety in a Late Medieval English Manuscript: London, British Library Additional 37049." In *History in the Comic Mode: Medieval Communities and the Matter of Person*, ed. Rachel Fulton and Bruce W. Holsinger, 182–91. New York: Columbia University Press, 2007.

———. "Morbid Devotions: Reading the Passion of Christ in a Late Medieval Miscellany, London, B.L. MS Additional 37049." Ph.D. diss., Columbia University, 2001.

Hillgarth, J. N. *Ramon Lull and Lullism in Fourteenth-Century France*. Oxford: Clarendon Press, 1971.

Hilton, Rodney. *Bond Men Made Free: Medieval Peasant Movements and the English Rising of 1381*. New York: Viking Press, 1973.

Hirsch, Rudolf. "The Invention of Printing and the Diffusion of Alchemical and Chemical Knowledge." In *The Printed Word: Its Impact and Diffusion*, 10:115–41. London: Variorum, 1978.

Holdenried, Anke. *The Sibyl and Her Scribes: Manuscripts and Interpretation of the Latin* Sibylla Tiburtina, *c. 1050–1500*. Aldershot: Ashgate, 2006.

Holmyard, Eric John. *Alchemy*. Harmondsworth: Penguin, 1957.

Jacquart, Danielle. "Medical Practice in Paris in the First Half of the Fourteenth Century." In *Practical Medicine from Salerno to the Black Death*, ed. Luis García-Ballester et al., 186–210. Cambridge: Cambridge University Press, 1994.

———. "Medical Scholasticism." In *Western Medical Thought from Antiquity to the Middle Ages*, ed. Mirko D. Grmek, 233–35. Cambridge, Mass.: Harvard University Press, 1998.

———. "Theory, Everyday Practice, and Three Fifteenth-Century Physicians." *Osiris* 2nd ser., 6 (1990): 140–60.

Jacquart, Danielle, and Claude Thomasset. *Sexuality and Medicine in the Middle Ages*. Trans. Matthew Adamson. Oxford: Polity Press, 1988.

Johnston, Mark D. *The Evangelical Rhetoric of Ramon Llull: Lay Learning and Piety in the Christian West Around 1300*. New York: Oxford University Press, 1996.

Jordan, William Chester. *The Great Famine: Northern Europe in the Early Fourteenth Century*. Princeton, N.J.: Princeton University Press, 1996.

Jullien de Pommerol, Marie-Henriette. "Les papes d'Avignon et leurs manuscrits." In *Livres et bibliothèques (XIIIe–XVe siècle)*, *Cahiers de Fanjeaux* 31 (1996): 133–56.

Kaup, Matthias. "Der *Liber Horoscopus*: Ein bildloser Übergang von der Diagrammatik zur Emblematik in der Tradition Joachims von Fiore." In *Die Bildwelt der Diagramme Joachims von Fiore: Zur Medialität religiös-politischer Programme im Mittelalter*, ed. Alexander Patschovsky, 147–84. Ostfildern: Jan Thorbecke Verlag, 2003.

Kelly, Samantha. *The New Solomon: Robert of Naples (1309–1343) and Fourteenth-Century Kingship* (Leiden: Brill, 2003).

Kermode, Frank. *The Sense of an Ending: Studies in the Theory of Fiction*. Oxford: Oxford University Press, 1966.

Kibre, Pearl. "Albertus Magnus on Alchemy." In *Albertus Magnus and the Sciences: Commemorative Essays 1980*, ed. James A. Weisheipl, O.P., 187–202. Toronto: Pontifical Institute of Mediaeval Studies, 1980.

Kieckhefer, Richard. *Unquiet Souls: Fourteenth-Century Saints and Their Religious Milieu*. Chicago: University of Chicago Press, 1984.

Kroll, Jerome, and Bernard Bachrach. "Sin and the Etiology of Disease in Pre-Crusade Europe." *Journal of the History of Medicine and Allied Sciences* 41 (1986): 395–414.

Kuhn, Thomas S. "Metaphor and Science." In *Metaphor and Thought*, ed. Andrew Ortony, 533–42. Cambridge: Cambridge University Press, 1993.

Lakoff, George, and Mark Johnson. *Metaphors We Live By*. Chicago: University of Chicago Press, 1980.

Lambert, Malcolm. *Franciscan Poverty: The Doctrine of the Absolute Poverty of Christ in the Franciscan Order, 1210–1323*. London: S.P.C.K., 1961.

———. *Medieval Heresy: Popular Movements from the Gregorian Reform to the Reformation*. 3rd ed. Oxford: Blackwell, 2002.

Landes, Richard. "Lest the Millennium Be Fulfilled: Apocalyptic Expectations and the Pattern of Western Chronography, 100–800." In *The Use and Abuse of Eschatology in the Middle Ages*, ed. Werner Verbeke, Daniel Verhelst, and Andries Welkenhuysen, 137–211. Leuven: Leuven University Press, 1988.

Lapidge, M. "The Stoic Inheritance." In *A History of Twelfth-Century Western Philosophy*, ed. Peter Dronke, 81–112. Cambridge: Cambridge University Press, 1988.

Lee, Harold, Marjorie Reeves, and Giulio Silano. *Western Mediterranean Prophecy: The School of Joachim of Fiore and the Fourteenth-Century Breviloquium*. Toronto: Pontifical Institute of Mediaeval Studies, 1989.

Leff, Gordon. *Heresy in the Later Middle Ages: The Relation of Heterodoxy to Dissent c. 1250–1450*. 2 vols. Manchester: Manchester University Press, 1967.

Lemay, Richard. *Abu Ma'shar and Latin Aristotelianism in the Twelfth Century*. Beirut: American University, 1962.

Lerner, Robert E. "Antichrists and Antichrist in Joachim of Fiore." *Speculum* 60 (1985): 553–70.

———. "The Black Death and Western Eschatological Mentalities." *American Historical Review* 86 (1981): 533–52.

———. "Ecstatic Dissent." *Speculum* 67 (1992): 33–57.

———. *The Feast of Saint Abraham: Medieval Millenarians and the Jews*. Philadelphia: University of Pennsylvania Press, 2000.

———. "Frederick II, Alive, Aloft, and Allayed in Franciscan-Joachite Eschatology." In *The Use and Abuse of Eschatology in the Middle Ages*, ed. Werner Verbeke, Daniel Verhelst, and Andries Welkenhuysen, 359–84. Leuven: Leuven University Press, 1988.

———. "Historical Introduction." In *Liber secretorum eventuum* by John of Rupescissa, ed. Robert E. Lerner and Christine Morerod-Fattebert, 13–85. Fribourg: Editions Universitaires, 1994.

———. "Medieval Millenarianism and Violence." In *Pace e guerra nel basso medioevo: Atti del XL convegno storico internazionale*, 37–52. Spoleto: Centro italiano di studi sull'alto medioevo, 2004.

———. "The Medieval Return to the Thousand-Year Sabbath." In *The Apocalypse in the Middle Ages*, ed. Bernard McGinn and Richard K. Emmerson, 51–71. Ithaca, N.Y.: Cornell University Press, 1992.

———. "Millénarisme littéral et vocation des juifs chez Jean de Roquetaillade." In *Textes prophétiques et la prophétie en Occident XIIe–XVIe siècle*, ed. André Vauchez, 21–25. Rome: École Française de Rome, 1990.

———. "Millennialism." In *Apocalypticism in Western History and Culture*, vol. 2 of *The Encyclopedia of Apocalypticism*, ed. Bernard McGinn, 326–60. New York: Continuum, 2000.

———. "The Pope and the Doctor." *Yale Review* 78 (1988–89): 62–79.

———. "'Popular Justice': Rupescissa in Hussite Bohemia." In *Eschatologie und Hussitismus*, ed. Alexander Patschovsky, František Šmahel, and Antonín Hrubý, 39–42. Prague: Historisches Institut, 1996.

———. "Refreshment of the Saints: The Time After Antichrist as a Station for Earthly Progress in Medieval Thought." *Traditio* 32 (1976): 97–144.

———. "Writing and Resistance Among Beguins of Languedoc and Catalonia." In *Heresy and Literacy, 1000–1530*, ed. Peter Biller and Anne Hudson, 186–204. Cambridge: Cambridge University Press, 1994.

Lewis, Warren. "Peter John Olivi: Author of the *Lectura super apocalipsim*: Was He Heretical?" In *Pierre de Jean Olivi 1248–1298: Pensée scolastique, dissidence spirituelle et société: actes du colloque de Narbonne, mars 1998*, ed. Alain Boureau and Sylvain Piron, 135–56. Paris: Vrin, 1999.

———. "Peter John Olivi: Prophet of the Year 2000. Ecclesiology and Eschatology in the Lectura Super Apocalipsim." 2 vols. Ph.D. diss., Tübingen, 1972.

L'Hermite-Leclercq, Paulette. "Le sang et le lait de la vierge." In *Le sang au moyen âge, Actes du quatrième colloque international de Montpellier*, 145–62. Montpellier: C.R.I.S.I.M.A., 1999.

Lindberg, David C. "Cosmology." In *Science in the Middle Ages*, ed. Lindberg, 288–90. Chicago: University of Chicago Press, 1978.

———. "Science as Handmaiden: Roger Bacon and the Patristic Tradition." *Isis* 78 (1987): 518–36.

Linden, Stanton J. *The Alchemy Reader: From Hermes Trismegistus to Isaac Newton*. Cambridge: Cambridge University Press, 2003.

———. *Darke Hierogliphicks: Alchemy in English Literature from Chaucer to the Restoration*. Lexington: University of Kentucky Press, 1996.

Little, Lester K. *Religious Poverty and the Profit Economy in Medieval Europe*. Ithaca, N.Y.: Cornell University Press, 1978.

Lochrie, Karma. *Covert Operations: The Medieval Uses of Secrecy*. Philadelphia: University of Pennsylvania Press, 1999.

Long, Pamela O. *Openness, Secrecy, Authorship: Technical Arts and the Culture of Knowledge from Antiquity to the Renaissance*. Baltimore, Md.: Johns Hopkins University Press, 2001.

MacCormac, Earl R. *Metaphor and Myth in Science and Religion*. Durham, N.C.: Duke University Press, 1976.

MacKinney, V. L. "*Dynamidia* in Medieval Medical Alchemy." *Isis* 24 (1936): 400–414.

Maclean, Ian. "Heterodoxy in Natural Philosophy and Medicine: Pietro Pomponazzi, Guglielmo Gratarolo, Girolamo Cardano." In *Heterodoxy in Early Modern Science and*

Religion, ed. John Brooke and Ian Maclean, 2–29. Oxford: Oxford University Press, 2006.

Marrow, James H. *Passion Iconography in Northern European Art of the Late Middle Ages and Early Renaissance: A Study of the Transformation of Sacred Metaphor Into Descriptive Narrative*. Kortrijk: Van Ghemmert, 1979.

Martin, Luther H. "A History of the Psychological Interpretation of Alchemy." *Ambix* 22 (1975): 10–20.

Matter, E. Ann. "The Apocalypse in Early Medieval Exegesis." In *The Apocalypse in the Middle Ages*, ed. Bernard McGinn and Richard K. Emmerson, 38–50. Ithaca, N.Y.: Cornell University Press, 1992.

McGinn, Bernard. "Angel Pope and Papal Antichrist." In *Apocalypticism in the Western Tradition*, 6:221–51. Aldershot: Variorum, 1994.

———. *Antichrist: Two Thousand Years of the Human Fascination with Evil*. San Francisco: Harper, 1994.

———. *Apocalyptic Spirituality*. New York: Paulist Press, 1979.

———. *The Calabrian Abbot: Joachim of Fiore in the History of Western Thought*. New York: Macmillan, 1985.

———. "Early Apocalypticism: The Ongoing Debate." In *Apocalypticism in the Western Tradition*, 1:2–39. Aldershot: Variorum, 1994.

———. "Introduction: John's Apocalypse and the Apocalyptic Mentality." In *The Apocalypse in the Middle Ages*, ed. Bernard McGinn and Richard K. Emmerson, 3–19. Ithaca, N.Y.: Cornell University Press, 1992.

———. "Joachim of Fiore's *Tertius Status*: Some Theological Appraisals." In *Apocalypticism in the Western Tradition*, 10:219–36. Aldershot: Variorum, 1994.

———. "*Pastor Angelicus*: Apocalyptic Myth and Political Hope in the Fourteenth Century." In *Apocalypticism in the Western Tradition*, 6:221–51. Aldershot: Variorum, 1994.

McVaugh, Michael R. "Alchemy in the *Chirurgia* of Teodorico Borgognoni." In *Alchimia e medicina nel medioevo*, ed. Chiara Crisciani and Agostino Paravicini Bagliani, 55–75. Florence: Sismel, 2003.

———. "Chemical Medicine in the Medical Writings of Arnau de Vilanova." *Arxiu de textos catalans antics* 23/24 (2005): 239–64.

———. "The *humidum radicale* in Thirteenth-Century Medicine." *Traditio* 30 (1974): 259–83.

———. *Medicine Before the Plague: Practitioners and Their Patients in the Crown of Aragon, 1285–1345*. Cambridge: Cambridge University Press, 1993.

———. "Moments of Inflection: The Careers of Arnau de Vilanova." In *Religion and Medicine in the Middle Ages*, ed. Peter Biller and Joseph Ziegler, 47–67. York: York Medieval Press, 2001.

Megivern, James J., C.M. *Concomitance and Communion: A Study in Eucharistic Doctrine and Practice*. Fribourg: University Press, 1963.

Mensa i Valls, J. *Arnau de Vilanova, espiritual: Guia bibliogràfica*. Barcelona: Institut d'Estudis Catalans, 1994.

Mews, Constant. "The World as Text: The Bible and the Book of Nature in Twelfth-

Century Theology." In *Scripture and Pluralism: Reading the Bible in the Religiously Plural Worlds of the Middle Ages and Renaissance*, ed. Thomas J. Heffernan and Thomas E. Burman, 95–122. Leiden: Brill, 2005.

Meyerson, Mark D., Daniel Thiery, and Oren Falk, eds. *'A Great Effusion of Blood': Interpreting Medieval Violence*. Toronto: University of Toronto Press, 2003.

Millet, Hélène. *Les successeurs du pape aux ours: Histoire d'un livre prophétique médiéval illustré*. Turnhout: Brepols, 2004.

Miskimin, Harry. *The Economy of Early Renaissance Europe, 1300–1460*. Englewood Cliffs, N.J.: Prentice-Hall, 1969.

Mollat, G. *The Popes at Avignon, 1305–1378*. New York: Harper and Row, 1965.

Moorman, John. *A History of the Franciscan Order: From Its Origins to the Year 1517*. Oxford: Clarendon Press, 1968.

Multhauf, Robert P. "John of Rupescissa and the Origin of Medical Chemistry." *Isis* 45 (1954): 359–67.

———. *The Origins of Chemistry*. 1966. Reprint, Langhorne, Penn.: Gordon and Breach Science Publishers, 1993.

———. "The Science of Matter." In *Science in the Middle Ages*, ed. David C. Lindberg, 369–90. Chicago: University of Chicago Press, 1978.

Needham, Joseph, with Lu Gwei-Djen. *Spagyrical Discovery and Invention: Magisteries of Gold and Immortality*. Vol. 5, pt. 2 of *Science and Civilisation in China*. Cambridge: Cambridge University Press, 1974.

Neugebauer, O. *Astronomy and History: Selected Essays*. New York: Springer-Verlag, 1983.

Newman, Barbara. *God and the Goddesses: Vision, Poetry, and Belief in the Middle Ages*. Philadelphia: University of Pennsylvania Press, 2003.

———. *Sister of Wisdom: St. Hildegard's Theology of the Feminine*. Berkeley: University of California Press, 1987.

Newman, William R. "The Alchemy of Roger Bacon and the *Tres Epistolae* attributed to him." In *Comprendre et maîtriser la nature au moyen âge: Mélanges d'histoire des sciences offerts à Guy Beaujouan*, Hautes études médiévales et modernes 73, 461–79. Geneva: Droz, 1994.

———. "'Decknamen or Pseudochemical Language'? Eirenaeus Philalethes and Carl Jung." *Revue d'histoire des sciences* 49 (1996): 159–88.

———. *Gehennical Fire: The Lives of George Starkey, an American Alchemist in the Scientific Revolution*. Cambridge, Mass.: Harvard University Press, 1994.

———. "The Homunculus and His Forebears: Wonders of Art and Nature." In *Natural Particulars: Nature and the Disciplines in Renaissance Europe*, ed. Anthony Grafton and Nancy Siraisi, 321–45. Cambridge, Mass.: MIT Press, 1999.

———. "An Overview of Roger Bacon's Alchemy." In *Roger Bacon and the Sciences: Commemorative Essays*, ed. Jeremiah Hackett, 317–36. Leiden: Brill, 1997.

———. "The Philosopher's Egg: Theory and Practice in the Alchemy of Roger Bacon." *Micrologus* 3 (1995): 75–101.

———. *Promethean Ambitions: Alchemy and the Quest to Perfect Nature*. Chicago: University of Chicago Press, 2004.

———. "Technology and Alchemical Debate in the Late Middle Ages." *Isis* 80 (1989): 423–45.

Newman, William R., and Anthony Grafton, eds. *Secrets of Nature: Astrology and Alchemy in Early Modern Europe.* Cambridge, Mass.: MIT Press, 2001.

———.. "Introduction: The Problematic Status of Astrology and Alchemy in Premodern Europe." In *Secrets of Nature: Astrology and Alchemy in Early Modern Europe*, ed. Newman and Grafton, 1–37. Cambridge, Mass.: MIT Press, 2001.

Newman, William R., and Lawrence M. Principe. "Alchemy vs. Chemistry: The Etymological Origins of a Historiographic Mistake." *Early Science and Medicine* 3 (1998): 32–65.

———. "Some Problems with the Historiography of Alchemy." In *Secrets of Nature: Astrology and Alchemy in Early Modern Europe*, ed. William R. Newman and Anthony Grafton, 385–431. Cambridge, Mass.: MIT Press, 2001.

North, John D. "Astrology and the Fortunes of Churches." *Centaurus* 24 (1980): 181–211.

———. "Celestial Influence—the Major Premiss of Astrology." In *'Astrologi hallucinati': Stars and the End of the World in Luther's Time*, ed. Paola Zambelli, 45–100. Berlin: Walter de Gruyter, 1986.

O'Boyle, Cornelius. *The Art of Medicine: Medical Teaching at the University of Paris, 1250–1400.* Leiden: Brill, 1998.

———. *Thirteenth- and Fourteenth-Century Copies of the* Ars medicine: *A Checklist and Contents Description of the Manuscripts.* Cambridge: Wellcome Unit for History of Medicine, 1998.

Obrist, Barbara. *Les débuts de l'imagerie alchimique XIVe–XVe siècles.* Paris: Sycomore, 1982.

———. "Les rapports d'analogie entre philosophie et alchimie médiévales." In *Alchimie et philosophie à la renaissance*, ed. Jean-Claude Margolin and Sylvain Matton, 43–64. Paris: Vrin, 1993.

———."Vers une histoire de l'alchimie médiévale." *Micrologus* 3 (1995): 3–43.

O'Leary, Stephen D. *Arguing the Apocalypse: A Theory of Millennial Rhetoric.* New York: Oxford University Press, 1994.

Ortony, Andrew, ed. *Metaphor and Thought.* Cambridge: Cambridge University Press, 1993.

Ottosson, Per-Gunnar. *Scholastic Medicine and Philosophy: A Study of Commentaries on Galen's* Tegni *(ca. 1300–1450).* Naples: Bibliopolis, 1984.

Ovitt, George. *The Restoration of Perfection: Labor and Technology in Medieval Culture.* New Brunswick, N.J.: Rutgers University Press, 1987.

Pagel, Walter. *Paracelsus: An Introduction to Philosophical Medicine in the Era of the Renaissance.* Basel: Karger, 1982.

Palmer, R. "Pharmacy in the Republic of Venice in the Sixteenth Century." In *The Medical Renaissance of the Sixteenth Century*, ed. A. Wear, R. K. French, and I. M. Lonie, 100–117. Cambridge: Cambridge University Press, 1985.

Paniagua, Juan A. *El maestro Arnau de Vilanova, médico.* Valencia: Catédra e Instituto de Historia de la Medicina, 1969.

———. "Notas en torno a los escritos de alquimia atribuidos a Arnau de Vilanova." *Archivo Iberoamericano de historia de la medicina* 11 (1959): 404–19.
Paravicini Bagliani, Agostino. *The Pope's Body*. Trans. David S. Peterson. Chicago: University of Chicago Press, 2000.
———. "Rajeunir au moyen âge: Roger Bacon et le mythe de la prolongation de la vie." *Revue médicale de la Suisse Romande* 106 (1986): 9–23.
———. "Ruggero Bacone e l'alchimia di lunga vita: Riflessioni sui testi." In *Alchimia e medicina nel medioevo*, ed. Chiara Crisciani and Agostino Paravicini Bagliani, 33–54. Florence: Sismel, 2003.
———. "Storia della scienza e storia della mentalità Ruggero Bacone, Bonifacio VIII e la teoria della 'prolongatio vitae.'" In *Aspetti della letteratura latina nel secolo XIII*, ed. Claudio Leonardi and Giovanni Orlandi, 243–80. Florence: La Nuova Italia, 1986.
Park, Katharine. *Doctors and Medicine in Early Renaissance Florence*. Princeton, N.J.: Princeton University Press, 1985.
Patterson, Lee. "Perpetual Motion: Alchemy and the Technology of the Self." *Studies in the Age of Chaucer* 15 (1993): 25–57.
Payne, Joseph Frank. "Arnald de Villa Nova on the Therapeutic Use of Human Blood." *Janus* 8 (1903): 432–35, 77–83.
Pedersen, Olaf. "Astronomy." In *Science in the Middle Ages*, ed. David C. Lindberg, 303–37. Chicago: University of Chicago Press, 1978.
Pelikan, Jaroslav. *The Growth of Medieval Theology 600–1300*. Vol. 3 of *The Christian Tradition: A History of the Development of Doctrine*. Chicago: University of Chicago Press, 1978.
Perarnau i Espelt, Josep. "L'*Allocutio christini* d'Arnau de Vilanova: edició i estudi del text." *Arxiu de Textos Catalans Antics* 11 (1992): 7–135.
———. "El text primitiu del *De mysterio cymbalorum ecclesiae* d'Arnau de Vilanova." *Arxiu de Textos Catalans Antics* 7/8 (1988–89): 7–182.
———. "La traducció catalana medieval del *Liber secretorum eventuum* de Joan de Rocatalhada." *Arxiu de Textos Catalans Antics* 17 (1998): 7–219.
———. "La traducció catalana resumida del *Vade mecum in tribulatione* (Ve ab mi en Tribulació) de Fra Joan de Rocatalhada." *Arxiu de Textos Catalans Antics* 12 (1993): 43–140.
Pereira, Michela. *The Alchemical Corpus Attributed to Raymund Llull*. London: Warburg Institute, 1989.
———. "Alchemy and the Use of Vernacular Languages in the Late Middle Ages." *Speculum* 74 (1999): 336–56.
———. "Arnaldo da Villanova e l'alchimia." *Arxiu de textos catalans antics* 14 (1995): 95–174.
———. "Heaven on Earth: From the *Tabula Smaragdina* to the Alchemical Fifth Essence." *Early Science and Medicine* 5, no. 2 (2000): 131–44.
———. "*Medicina* in the Alchemical Writings Attributed to Raimond Lull (Fourteenth–Seventeenth Centuries)." In *Alchemy and Chemistry in the Sixteenth and Seventeenth Centuries*, ed. Piyo Rattansi and Antonio Clericuzio, 1–16. Dordrecht: Kluwer Academic Publishers, 1994.

———. *L'oro dei filosofi: Saggio sulle idee di un alchimista del Trecento*. Spoleto: Centro Italiano di Studi sull'Alto Medioevo, 1992.

———. "Teorie dell'elixir nell'alchimia latina medievale." *Micrologus* 3 (1995): 103–48.

———. "Un tesoro inestimabile: Elixir e 'prolongatio vitae' nell'alchimia del '300.'" *Micrologus* 1 (1992): 161–87.

Pingree, David. *The Thousands of Abū Ma'shar*. London: Warburg Institute, 1968.

Pomian, Krzysztof. "Astrology as a Naturalistic Theology of History." In *'Astrologi hallucinati': Stars and the End of the World in Luther's Time*, ed. Paola Zambelli, 29–43. Berlin: Walter de Gruyter, 1986.

Potestà, Gian Luca. "Radical Apocalyptic Movements in the Late Middle Ages." In *Apocalypticism in Western History and Culture*, vol. 2 of *The Encyclopedia of Apocalypticism*, ed. Bernard McGinn, 110–42. New York: Continuum, 2000.

———. *Il tempo dell'Apocalisse: Vita di Gioacchino da Fiore*. Rome: Laterza, 2004.

Pouchelle, Marie-Christine. *The Body and Surgery in the Middle Ages*. Trans. Rosemary Morris. Oxford: Polity Press, 1990.

Preus, Anthony. "Galen's Criticism of Aristotle's Conception Theory." *Journal of the History of Biology* 10 (1977): 65–85.

Principe, Lawrence M. *The Aspiring Adept: Robert Boyle and His Alchemical Quest, Including Boyle's "Lost" Dialogue on the Transmutation of Metals*. Princeton, N.J.: Princeton University Press, 1998.

Reeves, Marjorie. *The Influence of Prophecy in the Later Middle Ages: A Study in Joachimism*. Oxford: Clarendon Press, 1969.

———. "Joachim of Fiore and the Images of the Apocalypse According to St. John." *Journal of the Warburg and Courtauld Institutes* 64 (2001): 281–95.

———. *Joachim of Fiore and the Prophetic Future: A Medieval Study in Historical Thinking*. London: S.P.C.K., 1976. Reprint, Stroud: Sutton, 1999.

Reeves, Marjorie, and Beatrice Hirsch-Reich. *The Figurae of Joachim of Fiore*. Oxford: Clarendon Press, 1972.

Rusconi, Roberto. "Antichrist and Antichrists." In *Apocalypticism in Western History and Culture*, vol. 2 of *The Encyclopedia of Apocalypticism*, ed. Bernard McGinn, 287–325. New York: Continuum, 2000.

Rutkin, H. Darrel. "Astrology, Natural Philosophy, and the History of Science, c. 1250–1700: Studies Toward an Interpretation of Giovanni Pico della Mirandola's *Disputationes adversus astrologiam divinatricem*." Ph.D. diss., Indiana University, 2002.

Ryan, Michael Alan. "That the Truth Be Known: Prophecy and Society in the Late Medieval Crown of Aragon." Ph.D. diss., University of Minnesota, 2005.

Ryan, W. F., and Charles B. Schmitt, eds. *Pseudo-Aristotle, The Secret of Secrets: Sources and Influence*. London: Warburg Institute, 1982.

Santi, Francesco. *Arnau de Vilanova: L'obra spiritual*. Valencia: Diputació Provincial de València, 1987.

Schabel, Chris. "Philosophy and Theology Across Cultures: Gersonides and Auriol on Divine Foreknowledge." *Speculum* 81, no. 4 (2006): 1092–1117

Scholz Williams, Gerhild, and Charles D. Gunnoe, eds. *Paracelsian Moments: Science, Medicine, and Astrology in Early Modern Europe*. Kirksville, Mo.: Truman State University Press, 2002.

Schwartz, Orit, and Robert E. Lerner. "Illuminated Propaganda: The Origins of the 'Ascende Calve' Pope Prophecies." *Journal of Medieval History* 20 (1994): 157–91.

Sidelko, Paul L. "The Condemnation of Roger Bacon." *Journal of Medieval History* 22 (1996): 69–81.

Singer, Dorothea Waley. "Alchemical Writings Attributed to Roger Bacon." *Speculum* 7 (1932): 80–86.

Siraisi, Nancy G. "Avicenna and the Teaching of Practical Medicine." In *Medicine and the Italian Universities*, ed. Nancy G. Siraisi, 63–78. Leiden: Brill, 2001.

———. *Medieval and Early Renaissance Medicine: An Introduction to Knowledge and Practice*. Chicago: University of Chicago Press, 1990.

———. "The Physician's Task: Medical Reputations in Humanist Collective Biographies." In *The Rational Arts of Living*, ed. A. C. Crombie and Nancy G. Siraisi, 105–33. Northampton, Mass.: Smith College Studies in History, 1987.

———. *Taddeo Alderotti and His Pupils*. Princeton, N.J.: Princeton University Press, 1981.

Sivin, Nathan. "Chinese Alchemy and the Manipulation of Time." *Isis* 67 (1976): 513–26.

Smith, Pamela H. *The Business of Alchemy: Science and Culture in the Holy Roman Empire*. Princeton, N.J.: Princeton University Press, 1994.

Smoller, Laura Ackerman. "The Alfonsine Tables and the End of the World: Astrology and Apocalyptic Calculation in the Later Middle Ages." In *The Devil, Heresy, and Witchcraft in the Middle Ages: Essays in Honor of Jeffrey B. Russell*, ed. Alberto Ferreiro, 211–39. Leiden: Brill, 1998.

———. "Astrology and the Sibyls: John of Legnano's *De adventu Christi* and the Natural Theology of the Later Middle Ages." *Science in Context* 20, no. 3 (2007): 423–50.

———. *History, Prophecy, and the Stars: The Christian Astrology of Pierre d'Ailly, 1350–1420*. Princeton, N.J.: Princeton University Press, 1994.

———. "Of Earthquakes, Hail, Frogs, and Geography: Plague and the Investigation of the Apocalypse in the Later Middle Ages." In *Last Things: Death and the Apocalypse in the Middle Ages*, ed. Caroline Walker Bynum and Paul Freedman, 156–87. Philadelphia: University of Pennsylvania Press, 2000.

Southern, R. W. "Aspects of the European Tradition of Historical Writing: 3. History as Prophecy." *Transactions of the Royal Historical Society* 5th ser., 22 (1972): 159–80.

Swanson, Robert N. "A Survey of Views on the Great Schism, c. 1395." *Archivum Historiae Pontificiae* 21 (1983): 79–103.

Sylla, Edith. "Autonomous and Handmaiden Science." In *The Cultural Context of Medieval Learning*, ed. John Emery Murdoch and Edith Dudley Sylla, 349–96. Dordrecht: Reidel, 1975.

Szulakowski, Urszula. *The Sacrificial Body and the Day of Doom: Alchemy and Apocalyptic Discourse in the Protestant Reformation*. Leiden: Brill, 2006.

Taylor, Frank Sherwood. *The Alchemists: Founders of Modern Chemistry*. New York: Schuman, 1949.

———. "The Idea of the Quintessence." In *Science, Medicine, and History: Essays on the Evolution of Scientific Thought and Medical Practice Written in Honour of Charles Singer*, ed. E. Ashworth Underwood, 247–65. London: Oxford University Press, 1953.

Temkin, Oswei. *Galenism: Rise and Decline of a Medical Philosophy*. Ithaca, N.Y.: Cornell University Press, 1973.

Theisen, Wilfrid, O.S.B. "The Attraction of Alchemy for Monks and Friars in the Thirteenth–Fourteenth Centuries." *American Benedictine Review* 46 (1995): 239–53.

Thijssen, J. M. M. H. *Censure and Heresy at the University of Paris, 1200–1400*. Philadelphia: University of Pennsylvania Press, 1998.

Thorndike, Lynn. *A History of Magic and Experimental Science*. 8 vols. New York: Columbia University Press, 1923–58.

———. *University Records and Life in the Middle Ages*. New York: Columbia University Press, 1944.

Torrell, Jean-Pierre. "La conception de la prophétie chez Jean de Roquetaillade." In *Textes prophétiques et la prophétie en Occident XIIe–XVIe siècle*, ed. André Vauchez, 267–86. Rome: École Française de Rome, 1990.

Vauchez, André, ed. *Textes prophétiques et la prophétie en Occident XIIe–XVIe siècle*. Rome: École Française de Rome, 1990.

Verrier, René. *Études sur Arnaud de Villeneuve v. 1240–1311*. 2 vols. Leiden: Brill, 1947.

Visser, Derk. *Apocalypse as Utopian Expectation (800–1500): The Apocalypse Commentary of Berengaudus of Ferrières and the Relationship Between Exegesis, Liturgy, and Iconography*. Leiden: Brill, 1996.

Ward, Benedicta. *Miracles and the Medieval Mind: Theory, Record, and Event, 1000–1215*. Philadelphia: University of Pennsylvania Press, 1982.

Webster, Charles, and Margaret Pelling. "Medical Practitioners." In *Health, Medicine, and Mortality in the Sixteenth Century*, ed. Charles Webster, 165–235. Cambridge: Cambridge University Press, 1979.

Weill-Parot, Nicolas. "Arnaud de Villeneuve et les relations possibles entre le sceau du Lion et l'alchimie." *Arxiu de textos catalans antics* 23/24 (2004–2005): 269–80.

Wilford, John Noble. "Transforming the Alchemists: Serious Scientists Revisit a Discredited Craft." *New York Times*, August 1, 2006.

Williams, Steven J. "Roger Bacon and His Edition of the Pseudo-Aristotelian *Secretum Secretorum*." *Speculum* 69 (1994): 57–73.

———. *The Secret of Secrets: The Scholarly Career of a Pseudo-Aristotelian Text in the Latin Middle Ages*. Ann Arbor: University of Michigan Press, 2003.

Wood, Charles T. "The Doctors' Dilemma: Sin, Salvation, and the Menstrual Cycle in Medieval Thought." *Speculum* 56 (1981): 710–27.

Yates, Frances A. "The Art of Ramon Lull." *Journal of the Warburg and Courtauld Institutes* 17 (1954): 115–73.

Zambelli, Paola, ed. *'Astrologi hallucinati': Stars and the End of the World in Luther's Time.* Berlin: Walter de Gruyter, 1986.

Zanier, Giancarlo. "Procedimenti farmacologici e pratiche chemioterapeutiche nel *De consideratione quintae essentiae.*" In *Alchimia e medicina nel medioevo*, ed. Chiara Crisciani and Agostino Paravicini Bagliani, 161–76. Florence: Sismel, 2003.

Ziegler, Joseph. "Alchemy in *Practica Summaria*: A Footnote to Michael McVaugh's Contribution." *Arxiu de textos catalans antics* 23/24 (2005): 265–67.

———. *Medicine and Religion, c. 1300: The Case of Arnau of Vilanova.* Oxford: Clarendon Press, 1998.

Ziegler, Philip. *The Black Death.* London: Collins, 1969.

Zupko, Jack. *John Buridan: Portrait of a Fourteenth-Century Arts Master.* Notre Dame, Ind.: University of Notre Dame Press, 2003.

Zygmunt, Joseph F. "Prophetic Failure and Chiliastic Identity: The Case of Jehovah's Witnesses." In *Expecting Armageddon: Essential Readings in Failed Prophecy*, ed. Jon R. Stone, 65–85. New York: Routledge, 2000.

Index

'Abū 'Abd Allāh al-Shī'ī, 134
Abū Ma'shar (Albumasar), 133–34, 136, 137, 213n. 2
Ad conditorem (John XXII), 23
Additiones (Giovanni d'Andrea), 90
aging, remedies against, 5, 64–65, 68–70, 83–84, 90, 145, 160
Alan of Lille, 146
Albert the Great, 56, 58, 71, 104, 186n. 12, 219n. 48; artifice, views of, 143–44
Alchemiae quam vocant (Gratarolo), 158
alchemy: apocalyptic, 8–9, 57–64; Arabic, 54–55; Chinese, 54, 56, 84; Christian, 4–6, 10, 99, 155; Greco-Egyptian, 54, 55, 56; interior spiritual development and, 105, 116; major concerns of, 54, 56; medieval, 54–55; natural vs. manmade products, 129–30; to oppose apocalyptic disasters, 52, 57–60, 99–101, 114–15, 145; parallel between human and nature, 35–36; recipes, 54–55, 62–63, 70–71, 75, 77–78, 93, 117, 119, 146, 159; resurrection of body and, 35, 38–39, 85–87, 107–10; revelation of secret processes, 59–60; rhetoric of, 2–3; Rupescissa healed by, 52; soul of alchemist, 115–16; sources, 52–57; as term, 63–64. *See also* medical alchemy; quintessence; transmutation of metals
alcohol. *See* aqua ardens
Aldebrandi, Stephen, 29
Alderotti, Taddeo, 105, 192n. 58
alembic, 113–14, 129
Alexander of Alexandria, 22
Alia informatio beguinorum (Arnald of Vilanova), 89
Allocutio christiani, 91
Alphidius, 111
analogies, 95–96
Andrew of Hungary, 131
Angelic Pope, 24, 25, 36, 37, 42, 115, 177n. 20
Angelo Clareno, 24, 171n. 34
animal stone, 119, 209n. 44
Anselm of Canterbury, 130–31
antediluvian myth, 84
Antichrist: alchemy used to oppose, 114–15, 145; appearance of, 2, 6, 44, 124, 132, 137; Arnald of Vilanova's view, 89–90; Bacon's view, 87; blood and, 125; eastern and western, 36; final, 17–18, 40, 177n. 18; French rebellion predicted, 49–50; human action, role of, 4, 6, 15–16, 33; identified with contemporary persons, 44–45; Llull's view, 96; multiple, 177n. 18; Nero, 36, 50; perfected health of evangelical men, 60–62, 65, 85, 87, 93, 107, 148–49; pope as, 22, 41; specific names and dates for, 2, 11, 22, 33–34, 44–45, 50, 156
antimony, 60, 117, 209n. 42
apocalyptic alchemy, 57–64

INDEX

apocalyptic crises, 13; alchemy and, 99–101; as allegorical, 15; mass conversions, 42–43; natural explanations for, 139–30; nature's role in, 7–8, 132–33, 140–41; need for specificity, 45–46; philosophers' stone used to oppose, 57–60, 114–15, 145; remedies for survival of, 50–51; specific dates for, 11–12, 33, 132–33, 148, 150–51, 156

apocalypticism, 2–3, 6, 8–9, 13, 99. *See also* Joachite prophecy; millennium; prophecies; Sibylline prophecy

aqua ardens (alcohol/burning water), 52, 62, 63, 70, 93–94; equated with fifth element, 71, 105–6

aqua vitae (water of life), 70, 94, 105–6, 159

Aquinas, Thomas, 63, 64, 135, 143, 145

Arabic alchemy, 54–55

Aragonese monarchy, 157

Arbor vitae crucifixae Jesu (Ubertino of Casale), 22

Archidoxis (Paracelsus), 160

Aristotelian natural philosophy, 7, 64, 83, 161, 163

Aristotle, 53, 56, 67, 143, 219n. 51; cosmology of, 66–67, 76

Arnald of Vilanova, 7, 22, 24, 64, 71–72, 79–81, 201–2n. 47, 216n. 18; blood imagery in works of, 125–26; cited by Rupescissa, 94, 112, 113; code names in works of, 104–5; ecstatic visions, 47; images of Christ's life in works of, 110; millennium prediction, 38, 39, 100, 136; nature, view of, 140; as physician, 89, 90; spiritual understanding and, 47, 62; vocations, view of, 91–92

ars combinatoria, 81, 95–96, 100

articella (*ars medicine*), 53–54

artifice/artificial, 86; nature vs., 129–30, 141–48, 219n. 51, 220n. 65

artisanal traditions, 5, 55

Ascende calve prophecy, 23–24

astrology, 7, 63, 73–79, 152, 193n. 64; celestial properties in quintessence, 73–74; Christian view of, 135; code names in, 72–73, 104; gold equated with sun, 75–77, 107; medicine and, 71–72; planetary-conjunction theory, 133–37, 214n. 10, 215n. 13, 216n. 18; Stoic pneuma, 67, 75; Zodiac Man, 130, *131*

astronomy, 104, 193n. 64

Atwood, Mary Anne, 105

Augustine, 14–15, 39–40, 46, 135, 139

Aurillac, prison at, 1, 24

Avicenna (al-Ḥusain b. 'Abdallāh Ibn Sīnā), 55, 56, 69, 142–43, 145, 218n. 40

Avignon papacy (Babylonian Captivity), 31, 32

Bacon, Roger, 7, 56, 64, 79, 80–81, 100, 127, 187n. 16; blood, view of, 118; equal complexion and, 83–87; Joachite influence on, 88–89; planetary-conjunction theory and, 134–36, 137, 216n. 18

Bagliani, Agostino Paravicini, 81

Balīnūs, 55

Bede, 17, 38

Beguins, 21–22, 23, 32, 41, 89, 171n. 36, 172n. 42

Benedictine order, 16

Benedict XI, 22

Bernard Gui, 22, 41

Bernard of Clairvaux, 120–22

Bibliotheca chemica curiosa (Manget), 158

Bignami-Odier, Jeanne, 32–33

Black Death, 2, 30, 72, 77–78, 133

blood: in devotional literature, 119–24; elixir of, 83; as fecund, 120–23; images of, 103, 116–27; miracle-host stories, 119, 122; preservative function of, 123–24; quintessence of, 73–74, 98, 119, 123, 142, 161; as symbol for violence, 124–25

blood cults, 122

bloodletting, 125

body: of Adam, 86; aging, 5, 64–65, 68–70, 83–84, 90, 145, 160; complexionary theory of, 67–70, 83–87,

{246}

96, 106–7, 145; humors, theory of, 63, 68, 106–7, 119, 125–26, 161; longevity of, 82–83, 92–93, 107, 197n. 8; as microcosm of world, 76–77; perfected, in third state, 100–101; purgation of, 5, 103, 126; resurrection of, 38–39, 83–87, 107–10
Bohemians, 157
Bonaventure, 19, 120–22, 135, 139
Boniface VIII, 22, 31, 82, 90
Book of Nature, 7–8, 129, 132, 138–39, 152
books of secrets, 154
Boyle, Robert, 162
Brown, Peter, 126
Brunschwygk, Hieronymous, 159
Buch der heiligen Dreifaltigkeit, 157–58
Bynum, Caroline Walker, 124

caelum humanum (human heaven), 94–95, 106–9, 128
Calvet, Antoine, 90, 92, 93
Camporesi, Piero, 119, 161
Canon of Medicine (Avicenna), 69
cardinals, flight of, 44
Carthusian miscellany, 120
Casteen, Elizabeth, 49
castration, 126
Catalan uprising, 35
Catherine of Siena, 126
Chartres, masters of, 220n. 65
chemistry, development of, 3, 161–62
chiliasm, 12, 13
Chinese alchemy, 54, 56, 84
Christ: blood of, 118–23, *121, 122*; images of, 7, 109–16; passion of, 110, 114, 117–18, 120, 129, 147; philosophers' stone identified with, 109–10, 113–15; as physician, 92; second coming of, 14; in visual art, 119–20, *121, 122*
Chronicon Moguntinum, 30
Church Fathers, 39, 48, 152
Ciompi rebellion (1378), 35
Cistercianism, 16
Clement IV, 82, 88, 135
Clement V, 22

Clement VI, 27, 43, 96
Clynn, John, 140
code names (*Decknamen*), 7, 76, 102, 112, 127–28; animal stone, 119, 209n. 44; in astrology, 72–73, 104; heaven, 105–9. *See also* imagery
Coelum philosophorum (Ulstad), 159
Commentum super prophetiam Cyrilli (Rupescissa), 28
Commentum super Veh mundo in centum annis (Rupescissa), 28–29
complexion: equal, 83–87, 96, 106–7, 145; of gold, 78
complexionary medicine, 67–68
Compositiones variae, 55
concordances (*concordiae*), 16–17, 129, 132, 137–39, 152–53
conjunction theory, 133–37, 214n. 10, 215n. 13, 216n. 18
Constantine of Pisa, 71, 72
Conventuals, 41
Correctio fatuorum, 90
cosmos (*mundi machina*), 43
Council of Constance, 158
Court, Guillaume, 32
Crisciani, Chiara, 74, 77–78
crucifixion, imagery of, 4, 7, 102–5, 110–11, 114, 120, 122–23, 153
Cum inter nonnullos (John XXII), 23, 42
Cum necatur flos ursi prophecy, 132–33, 137

da Bisticci, Lorenzo, 159
Daniel, E. Randolph, 89
De adventu Messiae (Llull), 96
De anima in arte alchimiae (Pseudo-Avicenna), 55, 119
De aqua ardente (Savonarola), 159
death, 64–65; of Christ, in alchemical imagery, 117–18; spontaneous generation, 111–12
De civitate Dei (Augustine), 39–40, 135
Decknamen. See code names
De congelatione et conglutinatione lapidum (Avicenna), 56, 142
De conservanda iuventute et retardanda senectute (Arnald of Vilanova), 90

{247}

De diversis artibus, 55
De generatione stellarum, 98
De humido radicali (Arnald of Vilanova), 90
De intentione medicorum (Arnald of Vilanova), 63
De iudiciis astronomie (Arnald of Vilanova), 72
De lapide philosophorum (*De secretis naturae*) (Arnald of Vilanova), 110
Della fisica (Fioravanti), 161
De longitudine et brevitate vitae (Aristotle), 67
De mineralibus (Albert the Great), 104
De mysterio cymbalorum (Arnald of Vilanova), 92, 110
De quinta essentia. See *Liber de consideratione quintae essentie omnium rerum*
De sanguine humano (attr. Arnald of Vilanova), 119
De sanitate tuenda (Galen), 68–69
De secretis nature (Llull), 9
Des monstres et prodiges (Paré), 160–61
De tempore adventus Antichristi (Arnald of Vilanova), 22, 38, 89, 92, 110
De virtutibus simplicium medicinarum (*Cogitanti mihi*), 62
devotional literature, 7, 43, 103, 116–17; images of blood in, 119–24
Dietrich of Arnevelde, 156
distillation, 3, 55, 152; of blood, 119; mercury-sulfur theory, 58–59; of philosophers' stone, 92; of quintessence, 58–59, 70–71, 92–94; of wine, 70, 78, 92–93, 98. See also quintessence
Dobbs, Betty Jo Teeter, 113
Dominican order, 41, 80
du Moulin, Antoine, 158
Durant, William, 90

earthquakes, 36, 140
Easton, Stewart, 88
ecstasy defense, 47, 182n. 69
Edward Prince of Wales, 30
Eiximenis, Francesc, 156–57
elements, 54, 55–56, 66; generation and, 82, 106; heavens as fifth element, 66–67, 69; in Pseudo-Llull, 97–98. See also fifth element
elixirs, 55–57, 78, 97, 119; equal complexion and, 83–87
embryology, 71
emperor, universal, 43, 44; Last World Emperor, 15, 36, 50
end of time/eschaton, 6, 13, 16–17, 21, 61, 97, 179n. 33, 220n. 4; exact dates of, 22, 44, 96; natural world and, 133–38, 145; perfection and, 109, 115; spiritual understanding and, 39–40, 48, 152. See also millennium
England, 157
English Peasant's Rebellion (1381), 35
equal complexion, 83–87, 96, 106–7, 145
eschatology, 12, 23, 31, 87, 96, 99–101, 103, 137–39. See also end of time/eschaton
Eternal Evangel, 16
Études sur Jean de Roquetaillade (Bignami-Odier), 32–33
Eucharist, 124, 210n. 54
evangelical men (*viri evangelici*), 76, 100–101, 152; health of, 60–62, 65, 85, 87, 93, 107, 148–49. See also spiritual men
exegesis, 40, 46–47; John of Rupescissa's, 4, 8, 29, 50, 100, 110–11, 113, 115, 118, 132, 139–40, 148; nature and, 132, 139–40, 148
Exiit qui seminat (Nicholas III), 42
extrascriptural revelations, 11–12

famine/Great Famine, 2, 10, 11, 30, 34, 36, 85, 150
Farinier, Guillaume, 26
Fatimid state, 134
Fernel, Jean, 70
Ficino, Marsilio, 8, 70
fifth element, 71, 75, 128; quintessence as, 66–67, 69–70, 73, 105–6. See also elements
Figeac, prison at, 24, 26
Fioravanti, Leonardo, 161
floral images, 120, 122
Fludd, Robert, 116

{ 248 }

fourteenth century, crises of, 1–2, 10, 13, 30–31, 35, 150–51, 154
France, 30–31, 36, 49
Franciscan order, 1, 80; Conventuals, 19–20, 41, 171n. 36; John of Rupescissa joins, 24; poverty, conflict over, 12, 18–23, 32, 40–43; in Rupiscessa's predictions, 36–37, 40–41, 115; Spirituals, 6, 20–22, 59, 89; tripartite division prediction, 41, 44, 171n. 36
Francis of Assisi, 18–19, 21
Fraticelli, 41, 171n. 36
Frederick II, 18, 181n. 57
Frederick III of Sicily, 91
Frederick of Brunswick, 156
free will, 135
Friedrich of Brandenburg, 158
Froissart, Jean, 30, 155

Galen, 54, 67–68
Galenic medicine, 7, 64, 68–69, 83, 117, 161, 163
Gaufré des Isnards, 82
generation, 82, 106, 111–12
Gentile da Foligno, 78
Genus nequam prophecy, 23–24
Gesner, Konrad, 159
Ghibellines, 36
Gilbert de la Porée, 146, 220n. 65
Giles of Rome, 64, 145
Giovanni d'Andrea, 90
Giustiniani chest, 161
Gloriosam ecclesiam (John XXII), 23
gold, 60, 62, 73, 80, 144–46; equated with sun, 75–77, 107; potable, 54, 77–78, 82, 84, 93, 195n. 83; quintessence of, 75–77, 98, 159
Grafton, Anthony, 72
Gratarolo, Guglielmo, 99, 158, 221n. 14
Great Schism of 1378, 156, 157
Great Year, 135
Greco-Egyptian alchemy, 54, 55, 56
Gregory IX, 19
Gruman, Gerald J., 84
Guelphs, 36

Haimo of Auxerre, 38
Halleux, Robert, 3, 5, 9
Hamburger, Jeffrey, 120, 124
heaven (*caelum*), 98; *caelum humanum* (human heaven), 94–95, 106–9, 128; heavens as fifth element, 66–67, 69; imagery of, 105–9; *nostrum caelum* (our heaven), 94, 106–9
Heinrich of Herford, 140
Hennessey, Marlene Villalobos, 120, 124
Henry of Harclay, 136, 140
Henry of Lancaster, 30
Henry of Rebdorf, 30
Hermes Trismegistus, 55, 104, 111
hierarchy: ecclesiastical, 21–22; social, 35–36, 43–44
Hildegard of Bingen, 17, 184n. 80
Hippocrates, 54, 68
history: ages of, 6, 12, 16–18, 20–21; of science, 5, 8; teleology of, 12–13, 129, 139, 149; trinitarian division of, 17, 21, 40, 171n. 30. *See also* salvation history; third state
Hitchcock, Ethan Allen, 105
Holmiensis papyrus, 54
Holmyard, Eric John, 116
Holy Roman Emperor, 36
Holy Spirit, 43
Honorius III, 19
host, 119, 122
Huernius, Joannes, 99
Hugh of St. Victor, 139
human agency, 100, 141, 148–52; in Joachite prophecy, 6, 16, 45, 49–50; power of, 4, 6, 12, 15–16, 33; prophecy and, 49–51
human heaven (*caelum humanum*), 94–95, 106–9, 128
humors, theory of, 63, 68, 106–7, 119, 125–26, 161
Hundred Years War, 2, 30–31
Hus, Jan, 157
Hussite uprising, 35, 157
hygiene, 65

{249}

iatrochemistry. *See* medical chemistry
imagery: alchemical, 7, 101, 103–5,
 117–18; of blood, 116–27; of Christ,
 7, 109–16; of crucifixion, 4, 7, 102–5,
 110–11, 114, 120, 122–23, 153; of
 heaven, 105–9; of resurrection, 7, 35,
 107–10. *See also* code names
Influence of Prophecy, The (Reeves), 45
influentia, 135
Informatio beguinorum seu Lectio Narbone
 (Arnald of Vilanova), 89
In libro sex scientarum in 3° gradu sapiente
 (Bacon), 82
Innocent IV, 19
Innocent VI, 30
Inquisition, 21–22, 23
Islam/Muslims, 43, 55, 95, 126, 134, 212n. 72
Israeli, Isaac, 62
Italy, 36, 37

Jābirian model, 84
Jābir ibn Hayyān, 54–55, 58
Jacquerie, 35
Jacquerie (uprising of 1358), 31
Jean de Murs, 43
Jean de Roquetaillade. *See* John of Rupescissa
Jean de Venette, 29–30
Jean le Bel, 30
Jehovah's Witnesses, 13
Jerome, 40
Jews, 15, 36, 43, 44, 91, 95, 134, 136–37, 156, 211n. 66
Joachim of Fiore: blurring of boundaries in, 108–9; *concordiae*, 16–17, 132, 138, 153; spiritual understanding of, 47, 62; third state, 6, 12, 16–18, 20; Trinity, doctrine of, 16–18, 21, 40, 96, 116, 171n. 30; *usus pauper* and, 20–21
Joachite prophecy, 5, 13, 38, 150, 153; Bacon, influence on, 88–89; human agency in, 6, 16, 45, 49–50; influence on Rupescissa, 32–33, 36, 87, 99–100
John I of Aragon, 157
John of Paris, 136, 140

John of Rupescissa: adaptation of prophecies of, 156–57; astrology and, 72, 104, 129; blurring of boundaries in works of, 108–9, 127–28, 138–39, 147–48, 153–55; claim to mystical authority, 1–2, 39–40, 46–47, 62–63, 65–66, 132; contemporary views of, 29–30; exegesis of, 4, 8, 29, 50, 100, 110–11, 113, 115, 118, 132, 139–40, 148; as *fantasticus*, 1, 27, 32, 96, 190n. 39; healed by alchemy, 52; hearing before papal curia, 1–2, 32; imprisonment, 1–2, 10, 23, 24–27; influence of, 3, 5, 8–10, 155–63; influences on alchemical theories of, 78–79, 80–81; Joachite influence on, 32–33, 36, 87, 99–100; joins Franciscan order, 24; on Llull, 95–96; optimism of, 34, 42, 51, 148, 151–52, 154; originality of, 6, 46, 80, 100, 151; pharmacological alchemy originated by, 53, 60; planetary-conjunction theory, 136–37, 214n. 10, 215n. 13; poverty and corruption, view of, 40–41; proving of, 23–31; theory, use of, 62–63, 111–12; trial of (1349), 27, 96, 140. *See also specific works*
John the Good, 30
John XXII, 21, 22, 23, 42
Judaism, 134
Jung, Carl, 105
Jungian models, 7, 105, 153

Kermode, Frank, 14
Kieckhefer, Richard, 126
Kitāb al-Shifā, 56
Konrad of Halberstadt, 29

language, apocalyptic tropes, 36–37. *See also* metaphors
last judgment, 13, 17, 38, 40
last things, 12, 14–16, 87
Last World Emperor, 15, 36, 50
Lectura super Apocalypsim (Olivi), 20
Lee, Harold, 89
Legat, Simon, 27–28

Leidensis papyrus, 54
Lerner, Robert E., 13, 33, 47, 156
Leviathan, 137
Liber contra Antichristum (Llull), 96
Liber de consideratione quintae essentie omnium rerum (Rupescissa), 2, 7, 9, 52, 57, 60–64; on artifice, 141–42, 146–47; on gold, 77, 145–46; imagery of blood in, 116–17, 119, 123, 142; manuscripts, 158, 221n. 14; quintessence described in, 64–65, 94, 106
Liber figurarum (Joachim of Fiore), 17
Liber lucis (Rupescissa), 2, 7, 9, 29, 52; alchemy used to oppose apocalyptic disasters, 57–60; Arnald, references to, 92, 112, 113; imagery of blood in, 116–17, 119, 123; imagery of Christ in, 110–11; manuscripts, 158
Liber ostensor quod adesse festinant tempora (Rupescissa), 1, 3, 24, 29, 30, 32, 43, 46, 132, 172n. 42, 184n. 80, 190n. 39
Liber principiorum medicine (Llull), 96
Liber secretorum eventuum (Rupescissa), 28, 33, 36, 60; Christ identified with philosophers' stone, 114; fate of Franciscans in, 41; French rebellion predicted, 49–50; human agency in, 148; innovation in, 46; Louis of Sicily as Antichrist, 44–45
Liber sex scientarum (Bacon), 85
Lindberg, David C., 89
Llull, Ramón, 7, 43, 59, 64, 79, 80–81, 95–99, 159; *ars combinatoria*, 81, 95–96, 100; code names in works of, 104; as *phantasticus*, 96
Lollard dissenters, 157
Louis of Siciliy, 44–45, 156, 181n. 57
Louis of Tarento, 131
Ludwig of Bavaria, 41
Luther, Martin, 116

Maineri, Maino de', 78
Manget, Jean-Jacques, 158
Mappae clavicula, 55
McGinn, Bernard, 48, 138
McVaugh, Michael R., 72, 91

medical alchemy, 52, 59–60, 71, 94; Black Death and, 77–78
medical chemistry, 3, 5, 60, 159
medical curriculum, 53–54
medical model for reform, 125–26
medicinal substances, 83; grades/degrees of, 67–68; potable gold, 54, 77–78, 82, 84, 93, 195n. 83
medicine: astrology as component of, 71–72; complexionary, 67–68; modern, influences on, 155–63; quintessence-based, 70–71; transcendent properties, 127–28
menstruation, 125, 212n. 68
menstruum (solvent), 98
mercury, 58–59, 60, 98, 104
mercury-sulfur theory, 58–59
Mesüe, John, 63
metaphors, 2–3, 7–8, 103–4. *See also* language, apocalyptic tropes
Meteorologica (Aristotle), 66, 142, 144, 219n. 51
Methodius of Patara, 15
Michaelists, 41, 42
Michael of Cesena, 23, 41
microcosm, 76–77, 147
Milk of the Virgin, 113, 117
millenarianism, 12, 13, 220n. 4
millenialism, 12, 13
millennium, 4, 17, 33; character of society in, 35–36, 43–44, 51; evangelical men, 60–62, 65, 85, 87, 93, 100–101, 103, 107, 111, 148–49; predictions of dates and duration, 38, 50, 151, 179n. 33, 220n. 4; resurrection of body and, 38–39, 85–86, 108–10; Rupiscessa's predictions, 36–44; understanding of scripture during, 39–40, 152
mineralogical theory, 77
miracles (*miracula*), 46–47, 63; categories of, 130–31; miracle-host stories, 119, 122
miraculous, quintessence as, 74–75
Multhauf, Robert P., 3, 5, 60, 187nn. 18, 19
Muslims. *See* Islam/Muslims

naturalism, 4, 6–7, 53, 60, 82, 96, 100, 103, 112, 127, 165n. 5
nature: apocalypticism within, 140–41; Book of Nature, 7–8, 129, 132, 138–39; divinity embedded in, 129, 155; end of time and, 133–38, 145; exact dates of apocalyptical events and, 132–33; exegesis and, 132, 139–40, 148; human mastery over, 145, 146; levels of, 73, 142; marvels, explanation of, 139–40; natural vs. artificial, 129–30, 141–48, 219n. 51, 220n. 65; reversal of roles in, 34–36; scripture identified with, 129, 130–41, 137–39. *See also* perfected nature
Nero (Antichrist), 36, 50
Newman, William R., 72, 102, 105, 143
Newton, Isaac, 162
Nicholas III, 42
nonnaturals, 65, 84
noster Mercurius (our Mercury), 98
nostrum caelum (our heaven), 94, 106–9

Obrist, Barbara, 104
Olivi, Peter, 12, 20–22, 32, 36, 38–39, 89, 181n. 53
On Christian Doctrine (Augustine), 46
On Generation and Corruption (Aristotle), 56
On Great Conjunctions (Abū Maʿshar), 133–34
On Minerals (Albert the Great), 58
On the City of God (Augustine), 14–15
On the Correction of Impostures (Paracelsus), 160
On the Heavens (Aristotle), 66
Opus trilogy (Bacon), 80, 82
Oracle of Cyril, 157
Order of Fiore, 16–18
Ordinem vestrum (Innocent IV), 19
Origen, 126
Otto of Taranto (Duke), 156

Paniagua, Juan A., 90
papal curia, 34, 81; Rupescissa's hearing before, 1–2, 32
Paracelsians, 5, 99, 161–63

Paracelsus, 60, 99, 159–60, 163
Paré, Ambroise, 160–61
Park, Katharine, 77
parousia, 14
Paul of Tarento, 56, 143, 145, 146, 147
peasant rebellions, 35
Pereira, Michela, 74, 78, 159
Pererii, Petrus, 29, 34
perfected nature, 7, 73–79, 85, 152; channeled to imperfect world, 73–74, 147–48; end of time and, 109, 115; human heaven, 94–95, 106–9; human intervention and, 142, 144, 148–49; third state and, 100–101. *See also* nature
Peter Lombard, 38
Petrus Bonus of Ferrara, 56, 71, 116, 143–44, 146; astronomical imagery in works of, 104; imagery of Christ in works of, 109–10
pharmacological alchemy, 53, 60
Philip IV, 31
Philip VI, 30
philosophers' stone, 54–55, 94, 187n. 16; Christ identified with, 109–10, 113–15, 117–18; colors of, 58–59, 117–18, 144, 209n. 42; gold and, 76–77; heaven equated with, 94, 106–9; mercury-sulfur theory, 58–59; to oppose apocalyptic disasters, 57–60, 115, 145; opposes Antichrist, 114–15; putrefaction of metal, 112–13. *See also* quintessence
Physica (Aristotle), 143
planetary-conjunction theory, 133–37, 214n. 10, 215n. 13, 216n. 18
pneuma, 67, 75
political and apocalyptic elements, 15, 33
political astrology, 134
popes: angelic, 24, 25, 36, 37, 42, 115, 177n. 20; Antichrist, 22, 41; Great Schism of 1378, 156; prophecies centered on, 23–24, 25; Repairer, 37, 42, 115, 149, 156
post-Antichrist tradition, 15–16, 17, 49
practica, 61–62

pre-Antichrist tradition, 15–16
Pretiosa margarita novella (Petrus Bonus of Ferrara), 56, 104, 109–10, 116, 144
prime matter, 55–56, 112, 119
Primer del Crestià (Eiximenis), 157
Principe, Lawrence M., 102
Pro conservanda sanitate (Vitalis of Furno), 81–82
prophecies: *Ascende calve*, 23–24; centered on papacy, 23–24; human agency and, 49–51; problem of, 44–48; Pseudo-Methodian, 15, 45, 49; Sibyilline, 13, 15–16, 45, 49, 50, 91, 184n. 80; Tiburtine Sibyl, 15–16, 184n. 80; visions as source of, 46–48. *See also* Joachite prophecy
Pseudo-Aristotle, 83, 118–19, 145
Pseudo-Arnald, 92–94, 127
Pseudo-Avicenna, 55, 117, 119
Pseudo-Geber, 56, 58
Pseudo-Llull, 95, 97–98, 158
Pseudo-Methodian prophecy, 15, 45, 49
Pseudo-Methodius, 15, 184n. 80
purgation, 5, 103, 126

Quaestiones quas quaesivit quidam frater minor (Llull), 96
Quaestiones tam essentiales quam accidentales (Pseudo-Arnald), 94
quintessence, 3, 7; of antimony, 117; artifice and, 141–42; of blood, 73–74, 98, 119, 123, 142, 161; celestial properties in, 73–74; class of chemicals, 60; complexionary theory and, 69, 106–7; described, 64–65; dichotomous nature of, 108; distillation of, 58–59, 70–71, 92–94; effect on body, 69–70; as fifth element, 66–67, 69–70, 73, 105–6; of gold, 75–77, 98, 159; idea of, 64–73; as incorruptible, 65–67, 147–48; influence on development of modern medicine, 155–63; as marvelous/miraculous, 74–75; as modern term, 162; as "our heaven"/human heaven, 76, 86, 94–95, 106–7, 128; in Pseudo-Llullian texts, 97–98; Rupescissa's theory as departure from Bacon, 84–85. *See also* alchemy; distillation; philosophers' stone
Quo elongati (Gregory IX), 19
Quorundam exigit (John XXII), 23

Raoul de Cornac, 27
rationalist approach, 91
Reeves, Marjorie, 21, 45, 88–89
regimen of health, 65, 82–83
Regula Bullata (Francis of Assisi), 19
Regula Prima (*Regula Non Bullata*) (Francis of Assisi), 18
relics, 119, 122
Renaissance, 8
Repairer, 37, 42, 115, 149, 156
resurrection, 35, 38–39, 83–87, 107–10, 118
Revelations (Pseudo-Methodius), 15
Reverendissime pater (Rupescissa), 49
Rhazes (Abū Bakr ibn Zakarīyā al-Rāzī), 55
Rieux, prison at, 26–27
Robert of Naples, 92
Roman Empire, 55
Roman vitriol, 58, 117
Rosarius philosophorum (attr. Arnald of Vilanova), 59, 84, 92, 112
Ruska, Julius, 102

safe houses, in caves, 50–51
St. Walburg, convent of, 120
saltpeter, 58
salvation history, 6–7, 12, 36, 45, 48–50, 100, 103; alchemy identified with, 127–29; contingency of, 49–50; nature and, 141, 147, 149. *See also* history
Sancta romana (John XXII), 23
Savonarola, Michele, 159
Saxon prophets, 156
scholasticism, 56, 130–31; *Concordantia*, 138; division of disciplines, 61–62
Sciant artifices (Avicenna), 142, 145
scientia experimentalis, 87–88
Scot, Michael, 72

scripture: Acts, 14, 16; Acts of John, 126; Book of Revelation, 139; clues to apocalypse in, 11–12, 14, 34; Daniel, 34, 38, 39, 114; Ezekiel, 38; Genesis, 146; Isaiah, 44, 113, 183n. 76; Job, 137; John, 111; Matthew, 126; nature identified with, 129, 130–41, 137–39; 1 Peter, 34; Psalm 117, 113; St. John's Apocalypse, 2, 14, 32, 36, 39, 138
Secret of Secrets (Pseudo-Aristotelian), 83, 118–19, 145
Sibylline prophecy, 13, 15–16, 45, 49, 50, 91, 184n. 80
Sicily, 91
Sigismund, Emperor, 158
Silano, Giulio, 89
Simon de Phares, 140
Siraisi, Nancy, 193n. 64, 195–96n. 86
Smoller, Laura, 136, 139–40
Solemnis Medicus, 78
Solet annuere (Honorius III), 19
soul, of alchemist, 115–16
Southern, R. W., 139
Speculum alchimiae (attr. Arnald of Vilanova), 112
Speculum iudiciale (Durant), 90
spiritual intellect, 47–48, 62, 152, 183n. 76
spiritual men, 20–22, 59, 61, 89, 94, 115. *See also* evangelical men
states, 17, 20. *See also* third state
Stoic pneuma, 67, 75
sublunary world, 75, 97, 108–9, 216n. 18
sulfur, 58, 104
Summa perfectionis (Pseudo-Geber), 58, 143
superlunary sphere, 7, 106
surgeons, 124–25
Sylvester, Pope, 38

Table Talk (Luther), 116
Taborites, 33
Tabula smaragdina (attr. Hermes Trismegistus), 55
Taylor, Frank Sherwood, 75
Telesphorus of Cosenza, 157
Tempier, Bishop, 136

Testament (Francis of Assisi), 19
Testamentum (attr. Llull), 59
texts: alchemical, 80, 92–94, 103, 153–54; nature as, 7–8, 129, 132, 138–39, 152; prophetic, 87–88; unity of, 139. *See also specific works*
Theatrum Chemicum (Zetzner), 158
theological doctrine, 7, 103
theorica, 61–62
theriac, 78, 196n. 88
Thesaurus (Gesner), 159
third state, 6, 12, 16–22, 36, 40, 85, 87, 181n. 53; blurring of boundaries in, 108–9, 147–49; perfected body and, 100–101; *usus pauper* and, 20–21. *See also* history; states
Thomas of Bologna, 78
Thorndike, Lynn, 90, 110
Tiburtine Sibyl, 15–16, 184n. 80
Tincture of Redness, 117–19
Toxites, Michael, 99
Tractatus de aqua vite simplici et composita (Pseudo-Arnald), 92–93
Tractatus de vinis (Pseudo-Arnald), 92–93
Tractatus parabolicus (Pseudo-Arnald), 92, 112
transmutation of metals, 52, 54–56, 152, 187n. 16, 218n. 40, 219n. 48; as artificial, 141–44; biological reproduction as metaphor for, 104; philosopher's stone and, 58–59; in Pseudo-Arnaldian texts, 94; putrefaction of metal, 112–13; quintessence and, 98; as transformation of human interior, 105, 116
tree of life, 86, 120, *121*
Trinity, doctrine of, 16–18, 21, 40, 96, 116, 171n. 30
triplicity, 133–34

Ubertino of Casale, 22, 24, 36, 45, 47, 157
Ulstad, Philipp, 159
understanding: increased during millennium, 39–40, 48, 152; prophecy compared with, 46–48; spiritual intellect, 47–48, 152, 183n. 76

universities, 53–54, 57, 64, 184n. 3, 185n. 5
University of Bologna, 53
University of Montpellier, 89
University of Paris, 53, 72
University of Toulouse, 61, 138
usus pauper (poor use), 20–21

Vade mecum in tribulatione (Rupescissa), 9, 11, 29, 30, 33–35, 60, 148; blurring of boundaries in, 108–9; Franciscan pope in, 36, 115; images of tribulation in, 87; manuscripts, 157, 176nn. 5, 6; predictions not derived from scripture, 44; third state in, 40, 108–9
van Broekhuizen, Daniel, 158
Venette, Jean de, 155
Verae alchemiae (Gratarolo), 158
Virgin Mary, 120
visions, 46–47

visual art, symbol of Christ's blood in, 119–20
vita apostolica, 18, 125
Vitalis of Furno, 81–82, 105
Vitis mystica, 120–22
vitriol (metallic sulfate), 60
von Lippmann, E. O., 102
Vos misistis (Rupescissa), 29

Weill-Parot, Nicolas, 90
Wenceslas of Prague, 158
William of Conches, 63
wine, distillations of, 70, 78, 92–93, 98
world soul, 75

Zayton (city), 24
Zetzner, Lazarus, 158
Ziegler, Joseph, 90, 91–92, 125–26
Zodiac Man, 130, *131*
Zygmunt, Joseph F., 13

GPSR Authorized Representative: Easy Access System Europe, Mustamäe tee 50, 10621 Tallinn, Estonia, gpsr.requests@easproject.com